Aquaculture Training Manual

Second Edition

Donald R. Swift
MSc

Fishing News Books

Fishing News Books
A division of Blackwell Scientific
  Publications Ltd
Editorial offices:
Osney Mead, Oxford OX2 0EL
25 John Street, London WC1N 2BL
23 Ainslie Place, Edinburgh EH3 6AJ
238 Main Street, Cambridge,
  MA 02142, USA
54 University Street, Carlton,
  Victoria 3053, Australia

Other Editorial Offices:
Librairie Arnette SA
2, rue Casimir-Delavigne
75006 Paris
France

Blackwell Wissenschafts-Verlag GmbH
Meinekestrasse 4
D-1000 Berlin 15
Germany

Blackwell MZV
Feldgasse 13
A-1238 Wien
Austria

First Edition published 1985
Second Edition published 1993

Set by Best-set Typesetter Ltd., Hong
Kong
Printed and bound in Great Britain by
Hartnolls Ltd, Bodmin, Cornwall

DISTRIBUTORS

Marston Book Services Ltd
PO Box 87
Oxford OX2 0DT
(*Orders*: Tel: 0865 240201
       Fax: 0865 721205
       Telex: 83355 MEDBOK G)

USA
Blackwell Scientific Publications, Inc
238 Main Street
Cambridge, MA 02142
(*Orders*: Tel: 800 759-6102
         617 876-7000)

Canada
Oxford University Press
70 Wynford Drive
Don Mills
Ontario M3C 1J9
(*Orders*: Tel: (416) 441-2941)

Australia
Blackwell Scientific Publications Pty
Ltd
54 University Street
Carlton, Victoria 3053
(*Orders*: Tel: (03) 347-5552)

British Library
Cataloguing in Publication Data
A Catalogue record for this book is
available from the British Library

ISBN 0-85238-194-8

Contents

Preface to the Second Edition

In the introduction to the first edition the hope was expressed that the *Aquaculture Training Manual* would prove to be of value to people training to work in aquaculture projects, particularly in developing countries. This has indeed proved to be the case and the demand for the book has led to the production of the second edition.

This second edition incorporates a number of fundamental changes to the structure of the manual. It was felt that the main sub-divisions would more clearly fall into three main parts instead of two: one on fisheries biology, the second on general husbandry techniques and the last on the production of individual species. This has meant some re-ordering of the material content of the manual. The chapter and sub-section headings have been set out in more detail and use has been made of a modern numbering system. An index has also been added.

Another fundamental change to be incorporated is the redefinition of the word 'prawn' in favour of 'shrimp'. Although dictionary definitions show that there are differences between prawns and shrimps, in practice there is considerable confusion in the terminology, which itself is based on ignorance and local preferences. In an attempt to standardize on the definitions the Food and Agricultural Organization of the United Nations has stated that it is preferable for the salt water penaeids to be called 'shrimp' and for freshwater penaeids to be called 'prawns'. This we have done.

In the first edition it was suggested that aquaculture could be thought of as originating with the production of food at the village level for local consumption, but with the later development of commercial aquaculture for the production of high-quality, high-cost products for national and international markets.

The trend towards commercial aquaculture has continued since the publication of the first edition with significant new developments, particularly in the farming of marine flat fish shrimp and salmonids. Recent years have seen the development of commercial aquaculture for tropical freshwater fish, particularly tilapia. This development has resulted in research work on the biology of the fish. These trends are well illustrated by the subjects covered in The International Symposium on Tilapia in Aquaculture held in Israel as long ago as 1983. The section on the biology of the fish covered a wide range of subjects including endocrinology and genetics. Farm management discussed the mass production of seed, cage culture and commercial tank culture – a long way from the original tilapia farming in earth ponds.

In marine aquaculture the great increase in salmon farming has brought its own rewards and problems; this is well illustrated in the Scottish salmon farming industry, where extensive development in cage culture of salmon in sea lochs has led to problems of disease and water pollution. Salmon ranching is now carried out in at least eleven countries in both the northern and southern hemispheres. Japan, North America and Russia together produce perhaps some 70 000 t per year by this method.

The success of commercial salmon farming has had the inevitable result of forcing down the market price of the product. Lower profits and increased problems must mean that the peak of salmon production cannot be far off; this will be especially true if farmers turn to other species of fish. This will become possible with the development of the commercial farming of fish like the turbot which is rapidly becoming an established industry with a potential initial production of 1000 tons per year.

One can prophesy that the next five years will see the rapid development of commercial farming of new species of marine fish, the continuing development of the commercial farming of tropical freshwater species and the continued development of semi-subsistence culturing. Other species which are now being either successfully reared (and are not covered in this manual) or are being intensively researched for their aquaculture potential include sturgeon, halibut, sea bass, sea bream, abalone, lobsters, crabs, crayfish and many ornamental fishes. There will therefore be an increasing need for a manual of this type to emphasise the importance of understanding the basic physiology and biology of farmed fishes together with their relationship to their environment. Without an informed work force and, in developing countries an informed team of extension workers, the possibility of serious problems developing on farms is greatly increased. This is well illustrated by the present state of salmon farming in Scotland where, in spite of a well-educated work force and a very high level of consultancy, problems of disease and pollution are currently of great concern.

Donald R. Swift

Introduction to
the First Edition

Aquaculture is the term used for the growing of aquatic animals and plants in fresh, brackish and sea water. Aquaculture has been practised for centuries, particularly in the countries of the Far East, where it has been carried on for over a thousand years. In the past it was something of a small-farmer industry where people produced food mainly for their own consumption or for that of their villages. Steadily increasing world population has, however, caused growing demand for protein throughout the world and led to more exploitation of the natural populations of aquatic animals.

Very early in his development, man became a farmer. In the case of aquatic animals, however, farming developed slowly, and at first the supply of aquatic animals to markets came primarily from fishing wild stocks. Recently this situation has been changing as the result of (1) the increasing costs of fishing due to rising prices of such items as fuel, equipment and wages, (2) overfishing in many areas due to growing demand and (3) depletion of many stocks due to pollution.

As a result wild aquatic animals have become more expensive and harder to obtain. In order to meet the shortfall in supplies many countries have turned to producing more by farming, so encouraging the development of modern aquaculture techniques and, in particular, largescale commercial aquaculture. Highly developed modern aquaculture, like modern terrestrial farming, is big business requiring large capital investment. It is therefore not surprising that the modern aquaculture industry concentrated on the production of highly-priced products in good demand: salmon, turbot, sole, trout, catfish, yellowtails, prawns and oysters.

The years 1960–85 saw a rapid increase in the commercial development of aquaculture; that is to say, operations which produce for sale and which depend on the sale of their products for their profit. At the same time there has been considerable growth in traditional semi-subsistence aquaculture; that is, aquaculture carried out by individual farmers or small communities, the products of which are either consumed by the families involved or are sold locally for immediate gain.

This was, of course, the original type of aquaculture. At its most simple farmers trapped wild stock, or collected and impounded wild stock left behind by receding flood waters or tides – a type of farming still widely practised today. Largely owing to the work of international agencies, this type of farming is gradually becoming more efficient, especially in the all-important aspect of seed production.

The production of any farm, large or small, is basically dependent on the supply of new stock to the farm; this new stock in aquaculture is called seed. If this supply is subject to the vagaries of nature, as is the case where farmers collect their seed from the wild, it cannot be totally reliable. If the supply fails for any reason then the farmers will be short of food for that year and commercial operations dependent on wild seed will have a harder time attracting capital for their ventures.

For this reason some of the most important aquacultural advances in recent times have been in the field of seed production in the farm, leading to closed-cycle farming systems much more under the control of the farmer. Closed-cycle farming systems make possible controlled breeding, leading to stock improvements, such as have taken place in terrestrial farming over the past two hundred years or so. Selective breeding makes it possible to develop those characteristics of an animal of most value to the farmer. Thus the developments which are brought about by research and development work sponsored by large commercial developers of aquaculture will also benefit non-industrial small-scale farmers. An almost exact parallel to this kind of development can be seen in modern chicken farming, where the developments achieved by the modern poultry industry are of direct benefit to the small-scale poultry producer.

It is important for any future farmer to understand at least the basis of the functional requirements of the animal that he wishes to farm. Good husbandry of any animal requires that the farmer understands the interplay between the animal and the environment in which it is kept; how the environment affects the animal and the basis of established aquaculture practices. Good husbandry reduces animal mortality and increases both growth rate and feeding efficiency. Good husbandry thus leads to higher production levels in both industrial and non-industrial aquaculture.

The main object of the manual is to provide basic background reading for people training in acquaculture, especially those in developing countries where people of disciplines other than biology are concerned with the development of aquaculture. It is hoped that the manual will also be of value to an industrialist who is approaching the subject for the first time and who wishes to have a broader appreciation of the operation.

Part 1
The Biology of Farmed Aquatic Animals

Prospective fish farmers should have a working knowledge of the biology of the main farmed species. Part 1 therefore commences with a general overview of the relationships between the more commonly farmed aquatic animals, then moves from a general survey of various aspects of fish biology to a more detailed, though still quite simple, view of the anatomy of fish, crustaceans and molluscs. As a further aid to understanding the biology of aquatic species, growth, nutrition and health are dealt with in the chapters that follow.

Chapter 1
Systematics

In order to have a better appreciation of the animals which are farmed in aquaculture it is helpful to understand their relationships to each other and to the other animals which are not farmed. To be able to do this it is necessary to look briefly at the way in which animals and plants are classified.

The study of the relationship of animals to each other and the naming of animals is called systematics. From a farmer's point of view the most important aspect of systematics is that it allows us to identify an animal we are farming, which we know by one name but which can be called by a different name elsewhere. In systematics the same animals all have the same scientific name. From a scientific viewpoint there is another important aspect of systematics, and this is that the classification of animals allows us to have a better understanding of the relationship of different types to each other, and indeed systematics is also an indication of the evolutionary relationship of living things.

Scientists have classified the whole of the animal and plant kingdom according to their anatomy and biology. This system of classification was started by Linnaeus in 1758, since when it has been refined and expanded by each generation of taxonomists, biologists who study the classification of animals and plants.

The following description of systematics and of the classification of the more common animals farmed in aquaculture is not intended to be an exhaustive coverage but merely a description to allow the farmer to appreciate how the animals receive their scientific names. It is worth while emphasizing the importance of this nomenclature to the farmer. Already workers with tilapia are beginning to realise that many so-called 'pure' lines of these fish are not pure and that there is a danger of some 'pure' lines being lost. This is very important as selective breeding to improve farm stock requires that pure lines of animals are available from which to obtain the various crosses which are of such value to farmers.

The basic unit of classification is the species. The definition of what is meant by species is the subject of much scientific discussion but, for the purpose of the manual, species can be defined as those animals which normally and successfully inter-breed. In nature many adaptations to prevent interspecific breeding are met; thus crosses between species are very rare in nature and even when they occur the result of the cross is nearly always infertile. Perhaps the best example of this is the mule. Species are in turn grouped into genera; genera into families; families into orders; orders into classes and classes into phyla.

The following classification of some of the more commonly farmed aquatic animals illustrates this classification system.

1.1 Fishes

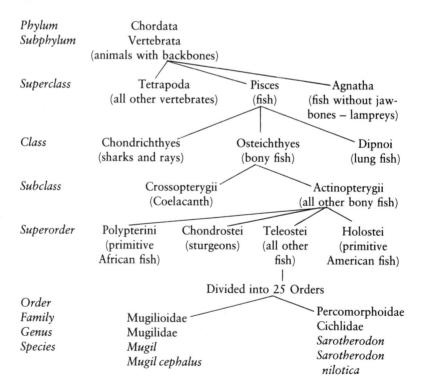

It can be seen from the above table that the vast majority of all fish are classified: Pisces – Osteichthyes – Actinopterygii – Teleostei. The Teleost fishes are then divided into 25 main groups called orders. These orders are divided into families, then genera and finally species. In the main a fishery biologist is concerned with orders of fish and downwards, which tells him all he needs to know about the fish he is dealing with.

1.2   Invertebrates (animals without backbones)

Shrimps and lobsters

Shrimps and lobsters belong to a subphylum of the phylum Arthropoda; the subphylum is called Crustacea. Crustacea largely consist of aquatic animals such as crabs, lobsters, shrimp, barnacles, water fleas, etc. However, the relationship of the commercially important crustaceans can be appreciated by starting at the level of order.

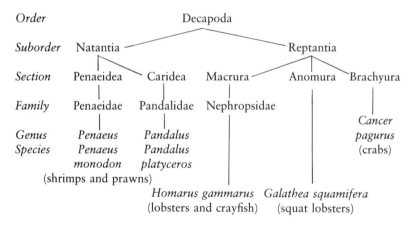

Molluscs

These are a large phylum of invertebrates, without segments but which usually have shells. The phylum Mollusca is divided into a number of classes:

Monoplacophora – deep sea molluscs
Aplacophora – primitive small molluscs

Polyplacophora – chitons
Scaphopoda – 'elephant tusk' molluscs
Gastropoda – slugs, snails and abalones
Cephalopoda – octopus and cuttlefish
Lamellibranchia (Bivalvia) – oysters and mussels

While there is considerable interest in many countries in farming abalones, most mollusc farming is of bivalves. Their relationship is as follows:

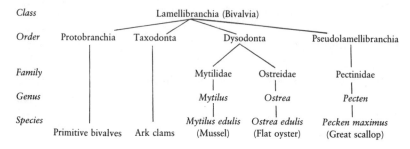

Chapter 2
General Biology of Farmed Aquatic Animals

2.1 Fish

The body shape and function of a fish is totally adapted to a free-swimming life in water, the body shape being adapted to give maximum efficiency to its propulsion through the water. Its internal organs, whilst performing the same basic functions as do those of a man, are adapted to perform those functions in water.

There are basically two shapes of fish, round fish and flat fish. Flat fish, such as skate, ray, plaice and sole are adapted to life on the bottom of the water body. Round fish, in general, have evolved an efficient hydrodynamic shape to allow them to move through the water body with the minimum expenditure of energy.

The animal's internal skeleton is used to form the frame to which are attached the muscles used for swimming and breathing. The fish propels itself through the water by sinusoidal movements of the body amplified by the flat tail. The fins attached to the side of the body are used for stability and to change direction in the same way as are the wings of a gliding bird.

The outer surface of the body is covered by skin. There are basically two types of fish skin, those covered in little bony plates called scales, and those without scales. Both types of skin secrete a mucus which makes the fish slippery to handle. This mucus has three main functions: it helps the passage of the fish through the water by reducing the friction of the skin; it helps to waterproof the fish; it gives protection against attacks of bacteria and other microscopic organisms which would otherwise infect the fish. It is therefore very important that fish are handled with care to avoid damage to the skin or its protective coating.

A fish breathes by extracting oxygen from the water in which it is living by means of gills. The gills are vascular organs situated behind the head and protected by a bony cover called an operculum. The gills function in much the same way as do our lungs; oxygen is extracted from the water, or from the air in the case of lungs, across a wet surface and into the blood stream of the animal. There is, however, one very important difference between air-breathing and water-breathing. In the air, except for changes due to large variations in altitude, the oxygen content stays more or less constant; in water this is not the case as the oxygen content of water decreases significantly with rising temperature. This is one of the basic facts of fish farming.

A land-living animal lives in an environment which is of a more or less constant chemical composition. This is not the case for aquatic animals. Water can vary considerably in its chemical composition, the extreme case being fresh water and sea water. Land animals live in constant danger of death by dehydration as the air in which they live contains much less water than do their bodies. Therefore, if they did not drink water and conserve water in their bodies they would die. The body loses water through evaporation and also through the production of urine.

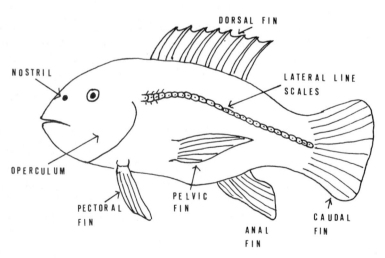

Fig. 2.1 General external features of a fish.

The kidneys produce urine in order to excrete unwanted substances from the body.

Fish are faced with two different situations depending on the salinity of the water in which they are living. Freshwater fish live in an environment which is much more dilute than is their body fluid. Therefore, as salt attracts water from the air so the fishes' body fluid attracts water and the fish live in constant danger of 'drowning' by an excessive intake of water into the body tissues. Seawater fish on the other hand live in water which is much more concentrated than is their body fluid and they face the danger of dehydration as water tends to leave their bodies.

Fish have similar sensory organs to man. They have eyes, but no eyelids; they have internal ears and balance organs but no external ear; they have no nose but are sensitive to the chemical content of the water and can 'smell' the chemicals. The organs of 'smell' are distributed over the skin but are also concentrated in pits in the head very similar to our nostrils. A fish has a well-developed nervous system with a brain and a spinal column, spinal nerves and cranial nerves. It seems very probable that a fish can experience some degree of pain although this is disputed by some scientists. However, as a fish has no vocal cords it is difficult to know!

A female fish lays eggs, which are fertilized by sperm produced by the male. The eggs and sperm in the great majority of species are released into the water in which the fish are living.

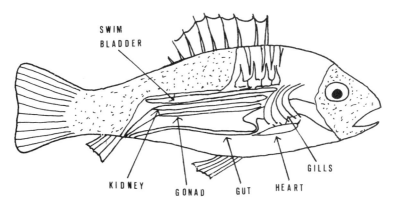

Fig. 2.2 Schematic general anatomy of a fish.

Fertilized eggs either float in the water body or fall to the bottom. Bottom eggs are sometimes guarded in nests by the fish or else they are sticky and become attached to weeds and stones; in some rare cases, including some tilapia which are farmed, the fish guard the eggs in their mouths. The eggs hatch in a few days or weeks depending on the species and the water temperature. The young fish hatch with small amounts of yolk attached to them which supplies them with the food they need during their first days of life. Some fish guard their young; some even guard them in their mouths. Other little fish have to fend for themselves.

Fish catch their food in their mouths, which are equipped with teeth. The food is not, however, chewed, although some plant-eating species have grinding plates in the mouth cavity which are used to macerate the vegetable matter eaten by the fish. The food is swallowed and digested in an alimentary canal similar to ours.

One very important difference between fish and mammals is that fish are cold blooded and so their body temperature is always the same as that of the water in which they are living. As body temperature plays a very important role in determining such things as growth rate, this fact is of great importance to fish farmers.

2.2   Crustaceans

The various problems encountered by a fish living in the sea, such as water temperature, levels of dissolved oxygen, and the high salt content of sea water, are all problems for crustaceans.

The most immediately obvious thing about a crustacean is the fact that it is covered on the outside by a hard surface, which is hardest in the crabs and lobsters. This hard outer shell is in fact continuous over the whole of the animal and is formed by a series of hard rings which are jointed together by a thin flexible outer layer. This hard outer layer creates one obvious problem for the animal, namely how to grow in size. The crustacean solves this problem by periodically casting the whole of its outer hard layer. The shell is cast and at this stage the animal is covered by a soft flexible skin. During this period the animal absorbs water and grows rapidly. After it has swelled the shell is

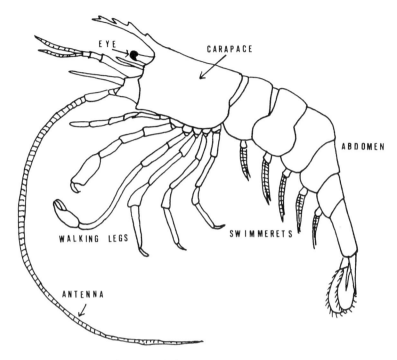

Fig. 2.3 General external features of a decapod crustacean.

hardened by the deposition of calcium salts so that in a few hours the animal has regained its former state but is larger. During the young stages of the animal these moults occur very frequently and often at the moult the animal changes its form. This change is called a metamorphosis.

Another unusual feature of crustaceans which is of great value to the animal is its ability to cast a limb if it is caught by it. Afterwards the animal can regrow the limb in stages at each moult.

The Decapoda – lobsters, crabs and shrimps – have a simple alimentary canal consisting of a straight tube with two anterior chambers. They have a simple blood system, a rudimentary heart and simple nervous and excretory systems.

The sexes are separate, the male fertilizing the female, which stores the sperm and releases them on laying her eggs. Crustaceans have a complicated life cycle during which they pass from the egg stage to the adult form stage through many different shapes and larval forms.

2.3 Molluscs

Bivalve molluscs live inside a hard shell which can be closed tight to give complete outer protection. Another feature is that most are immotile. One group, the scallops, can swim by clapping the two halves of the shell, but most of the important commercial species do not move. Mussels are secured to where they are living by special anchoring threads and oysters are fastened by a cement produced by the animal when it first settles. The bivalves have a simple digestive system and a simple heart, nervous system and gonads. Some species even have 'eyes', small light-sensitive organs.

The most outstanding features of the molluscs' anatomy are large organs looking like the gills of a fish. These are in fact the animal's gills and they act as a respiratory organ. They also act as an elaborate filter system which extracts microscopic particles of food from the water that the animal pumps through

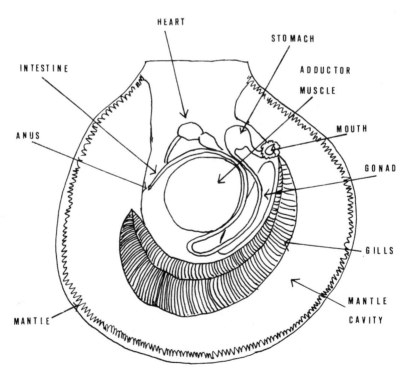

Fig. 2.4 Schematic general anatomy of a bivalve mollusc.

them. Later in this manual the importance of plants as the primary producers of food by the use of energy from sunlight is explained. Of all the animals farmed oysters, mussels and scallops are probably the truest vegetarians, needing little if any animal protein in their diet in order to survive and grow. Thus these molluscs which filter their feed out of the great volumes of sea water pumped through their gills are able to exploit very large areas of sea and so achieve very high production levels per hectare.

The bivalves have gonads which produce eggs and sperm. The eggs are fertilized in the water into which the eggs and sperm are released. The European oyster is an exception to this rule in that the eggs are retained inside the shell of the animal until hatching. The egg hatches into a free swimming larva which eventually settles and metamorphoses into the adult form.

Chapter 3
Anatomy

3.1 Fish

Skeletal and muscular system

Skeletons of fish are of two types, cartilaginous and bony. Cartilaginous skeletons never become calcified and therefore remain soft. These are the types of skeletons which skates, rays and sharks have. All other fish lay down calcium salts in their skeletons to form bone of varying degrees of hardness.

The function of the skeleton is to form a physical support for the body and a base against which the swimming muscles can pull. As the fish is living in water the body is supported by the water and so the function of the skeleton, unlike that of a land animal, does not include the necessity to support the body of the animal against the force of gravity.

The skeleton consists of a skull case which protects the brain, fastened to which is a flexible spinal column built up of vertebrae (Fig. 3.1). From the backbone there are bones, a little like our ribs, against which the powerful swimming muscles can pull. Most of the flesh of the fish consists of these muscles which are located on either side of the vertebral column. The muscles are alternately flexed and contracted so as to cause the fish's tail to move from side to side and thus to push it forward. There are bones which support the fins and the tail and various small bones in the muscles which aid the function of the muscles. Attached to the skull are jaws. Behind the skull is a bony plate called the operculum which protects the gills and also, when moved open and closed, helps pump water over the gills.

The skin of most fish is covered, to a greater or lesser extent, with small bony plates called scales. These plates are not, as is commonly supposed, outside the skin but are embedded in the

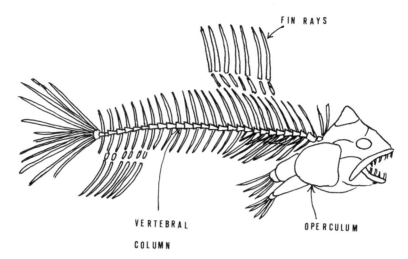

Fig. 3.1 Fish skeleton (schematic).

skin and so cause damage if they are rubbed off. The scales grow in size as the fish grows. The scales tend to grow at different rates at different times of the year varying with the changing growth rate of the fish. Changes in the growth pattern of the scales are marked by rings in the scale structure. Because of these marks the scales can be used to determine the age of the fish.

Circulatory system

In the section on fish nutrition it is explained that in order to release the energy contained in its food a fish requires oxygen. The oxygen for this process is extracted from the water by the gills and is transported from the gills to the tissues by the blood. In the blood the oxygen is carried in combination with a red chemical containing iron called haemoglobin which is carried through the blood in red blood cells. The blood is directed round the body in the blood vessels through which it is pumped by the heart.

In man the heart has four chambers; the blood passes from the lungs to the heart and is then pumped from the heart to the body. In a fish the heart is much more simple and has only two chambers, which are 'in line' in the main blood vessel and thus act as a straight pump. The blood returns to the heart from the body, is pumped to the gills for oxygenation and then continues on to the body.

The gills are situated at the side of the head behind the eyes and are covered by the opercular plate. The gills are feathery structures, very thin-skinned and containing many blood vessels. As the water is passed over the gills carbon dioxide, a product of the metabolism of the fish, diffuses from the blood into the water and oxygen diffuses from the water into the blood. Water is actively pumped over the gills by the fish using the muscles of the mouth and the operculum.

Digestive system

The gut is a tube which stretches from the mouth to the anus. It is divided into four parts:

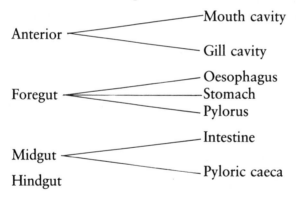

Anterior

The mouth cavity and jaws are used for capturing or taking up food. There are teeth around the jaw edge and backward-facing teeth inside the mouth; these are very well developed in fish which eat other fish to enable them to hold the prey. Inside the mouth and on the bars which support the gills are grinding plates; these are well developed in fish which eat vegetable matter.

Foregut

The stomach varies considerably in size and shape depending on the type of food eaten. For example, continuous mud-eaters have hardly any stomach whilst fish-eaters have large stomachs in which to retain their prey for initial digestion. The passage of the food from the stomach to the pylorus is controlled by a muscular ring called the pyloric sphincter.

Midgut

The digestive processes of the midgut have not been extensively studied; it is thought that the pyloric caeca have no special function but apparently merely serve to enlarge the surface area of the gut.

Hindgut

The hindgut continues the process of water and food absorption. The total length of the gut differs according to species, depending on the type of food they eat. Thus detritus-eaters have the longest gut and carnivores the shortest.

Digestion

The processes of digestion have been little studied and the times of digestion are not well known. As a fish is cold-blooded, times of digestion are much longer than in a warm-blooded animal and it may take days for a fish to digest a meal. There does not appear to be any digestive enzyme produced in the mouth of a fish and digestion appears to start in the stomach. Digestive enzymes are chemicals produced by the body which are passed into the lumen of the gut, where they break down food into its component parts, which can then be absorbed through the gut wall into the blood stream. In a fish the stomach produces an enzyme called pepsin which digests protein. The midgut produces trypsin which further digests proteins, lipase which digests fats and amylases which digest carbohydrates.

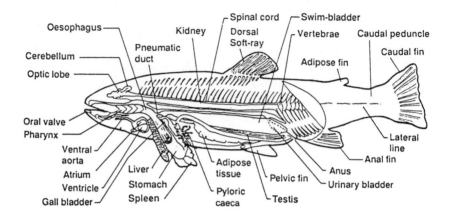

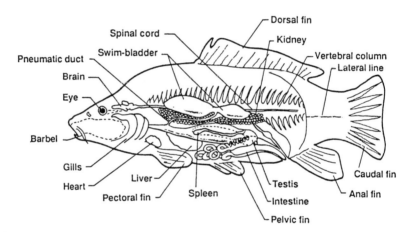

Fig. 3.2 Organ systems of trout (upper) and carp (lower) (from Shepherd & Bromage: *Intensive Fish Farming*).

Liver

The liver appears to function in the same way as in other vertebrate animals. It is the chemical factory of the body. Here waste products of the body are processed for excretion and the products of digestion are processed for use. As in other vertebrates the liver discharges a liquid into the gut called bile which contains some waste products from the body. It also emulsifies fats so that they can be digested more easily. The active components of bile are bilirubin and biliverdin which are the products of the breakdown of haemoglobin, the red pigment of the blood.

Excretory system

The products of the functioning of the fish's body and the products of the 'burning' of food to produce energy are got rid of by the body in two ways, via the respiratory system and via the excretory system. The kidneys, which are the main organs of the excretory system, are large and are situated along the top of the abdominal cavity.

The kidney has two important functions. It excretes waste products from the body and it plays an important role in maintaining the equilibrium of the body fluids. Freshwater fish take up water from their environment whilst sea fish lose water to their environment. In order to maintain the composition of their body fluids freshwater fish excrete a copious dilute urine, the kidneys retaining the dissolved salts within the body. Fish living in the sea have to drink water in order to survive since they are in constant danger of dehydration as they lose body water. Drinking water allows the fish to maintain its water content but it means that it also takes in unwanted levels of sea salts. A marine fish gets rid of unwanted chloride via its gills and its kidney excretes a urine equal in salt content to the sea water. Some fish, such as salmon, migrate from sea to fresh water and have kidneys which can adapt to either condition.

The kidney of a fish is not as well developed as is that of a land vertebrate. One result of this more primitive kidney is that a fish cannot excrete various nitrogenous compounds which a land vertebrate can. These compounds are deposited under the skin in the form of a chemical substance called guanine. Guanine reflects the light and so appears silvery, hence the silver colour of a fish.

Endocrine organs

In a vertebrate animal the awareness of the outside world is transmitted to the brain via the sense organs of touch, smell, taste, sight, temperature and vibrations. These sensations are transmitted to the brain by the nerves and are immediately reacted to in various ways. The animal also reacts to its environment in a slower way, this reaction being brought about by the secretion of chemicals into the body by endocrine organs. These chemicals are called hormones and affect the long-term reaction and development of the animal, sexual maturity and growth being examples.

A fish has nearly the same endocrine organs as other vertebrates. Obviously their function is of great importance to the fish farmer.

Pituitary

This small gland, located underneath the brain has been described as the 'master gland' of the body. The hormones produced by the pituitary affect the fish directly but also it produces hormones which act via the other endocrine organs. Thus the pituitary, which receives its signals direct from the brain, controls the reaction of the other endocrine organs to the environment and hence to a greater or lesser extent the actual activity of those organs. The pituitary produces the growth hormone which is believed to have a direct effect on the growth of the body as it has in many vertebrates. Nevertheless, its exact function is not fully understood. Obviously it is important for farmers to be aware of the growth hormone.

Thyroid

The thyroid is situated under the gills along the main blood vessel which runs between the heart and the gills. In man the thyroid controls the rate of metabolism, the rate at which the engine runs, in fact. The function of the thyroid in fish is not yet fully clear although there is some evidence to suggest that it also has an effect on the metabolism.

Gonads

Fish are either male or female and as such have either testes or ovaries. The gonads lie in the body cavity. Those of many marine fish are eaten as roe and the sturgeon's form the much-prized caviar. As well as producing eggs or sperm the gonads also secrete hormones into the blood stream. These hormones are responsible for the differences in appearance and behaviour between the males and females. The ripening of the gonads and the spawning activities of the fish which result from the hormones secreted into the blood stream by the gonads are, in the main, the result of changes in the fish's external environment. These changes act via the sense organs to the brain, from the brain to the pituitary and from the pituitary via hormones to the gonads. In this way changes outside a fish cause changes inside it which in turn can cause changes in activity.

A fish egg is composed of a tough outer shell, the embryo and a yolk store for the nutrition of the developing fish. The size of the egg and the amount of yolk vary with the species. In most tropical species the female lays many small eggs, perhaps 1 mm in diameter, which quickly develop and hatch within a few days of being laid. Salmon and trout, on the other hand, lay fewer but larger eggs, about 5 mm in diameter with larger yolk stores. These eggs take longer to hatch and the larvae have large yolk sacs attached to them on hatching.

Application of hormones in farming

In recent years two very important applications of hormones have been developed in fish farming.

The first concerns the use of pituitary hormones to induce a fish to spawn. It has been found in practice that many fish, such as the mullet and grass carp, which are of great importance in farming, will not spawn naturally under farm conditions. Grass carp, for example, need to complete long migrations before their gonads ripen. Obviously this is an effect of the external environment on the physiological state of the fish. However it has been discovered that if adult fish nearing maturation are given injections of pituitaries taken from other ripe fish the recipient fish will become ripe and will spawn under farm conditions. This procedure has been further refined by the use of injections of hormones and not whole pituitary extracts.

The second use of hormones concerns the control of population growth in tilapia which by excessive breeding can result in farm ponds being superpopulated with small fish of low market value. Obviously one way of controlling this would be to rear unisex populations. Various ways have been tried to obtain such populations and recently a simple effective system has been developed. This technique depends entirely on the use of the hormones produced by the gonads to control the sex of the fish. It has been found that if very young fish are dosed with male sex hormones an all-male population results, and this technique is now being successfully utilized with various species.

Nervous system and sensory systems

Fish have a well-developed nervous system with a complicated well-developed brain and spinal cord. There are nerves from the spinal cord to the body as well as nerves directly from the brain which perform special functions. The nervous system is in fact comparable to that of other vertebrates.

The sense organs of an animal are those parts of the body which allow the animal to be aware of its external environment and changes in that environment. These organs can be classified according to their function as follows: photoreceptors – sensitive to light; chemo and thermoreceptors – sensitive to changes in water chemistry and temperatures; mechano-receptors – sensitive to touch and vibration.

Photoreceptors

Fish have a pair of eyes which are constructed in the general plan of vertebrate eyes. That is to say they have a transparent frontal cornea, and an internal lens which focuses the received image onto a light-sensitive retina at the back of the eye. The eye can be turned by eye muscles to allow the fish to follow images. The focusing mechanism of the fishes' eye is different to that of man in that little change in the shape of the lens occurs but, in order to focus, the lens is moved in respect of the retina.

Chemoreceptors

A fish has three types of chemoreceptors, an olfactory organ with which it smells, taste buds, and a general set of sensory cells which are scattered over the skin all over the body with which it detects changes in the water chemistry.

The olfactory organs are situated in closed pits on the upper side of the head in front of the eyes. In sharks, which have a keen sense of smell for hunting their prey, the pit is open to the mouth and the water current passes through it. In other fish the pits are closed to the mouth but generally have two openings to the exterior so that the water passes through the pit.

The taste organs are situated in the mouth and also, except in sharks, similar organs are situated all over the fish's body. The general sensitivity of the fish to changes in the chemistry of the water is also due to a scattering of nerve endings which occur all over the body surface. A fish is therefore very sensitive to the chemistry of the water in which it is living and can detect very small changes in that chemistry. Salmon fry, for example, can detect some chemicals at a dilution of 1×10^{-11}. This sensitivity to dissolved substances is of importance to the fish farmer who must always remember the awareness of fish to chemical changes in the water.

Thermoreceptors

Fish are very sensitive to changes in water temperature, again important to remember as farmers may needlessly subject fish to sudden changes in water temperature. The temperature receptors are situated in the skin and are scattered over the body surface. It has been demonstrated that fish can be sensitive to changes in temperature as low as 0.03 °C.

Mechanoreceptors

Mechanoreceptors include organs sensitive to sound waves and vibrations, including water movements. A fish has two types of mechanoreceptors, the inner ear and the lateral line.

A fish possesses an inner ear on each side of its head similar in

structure to the inner ear of man. These structures are sensitive
to vibrations of the frequency of sound waves travelling through
the water.

The second system is called the lateral line system. This runs
down the length of the body, half way up the side. The lateral
line is a canal-like structure situated in the skin, containing
sensory cells equipped with hairs which project out into the
canal. The hairs bend in water currents and hence are sensitive
to vibrations and to water movements.

A fish farmer should be aware that his animals are very
sensitive to sound and to vibrations. Indeed, as sound travels
five times faster in water than it does in air, the fish in a pond
may hear a sound before the farmer.

The inner ear, as in other vertebrates, consists of semicircular
canals connected to a central chamber. In the central chamber is
a small flat bone called an otolith. As is explained later, under
ageing of fish, this otolith is used to determine age. The inner ear
lining contains sensory hair cells which detect sound vibratons,
head movements and also the position of the fish relative to the
pull of gravity. This organ of balance is therefore important to a
fish for it is often the case that a fish in deep or muddy water has
no reference point when it cannot see the bottom. Without the
inner ear it would not know which way up it was.

3.2   Crustaceans

Skeletal and muscular system

Crustaceans have no internal skeleton like that of a fish but they
have evolved an external one, called an exoskeleton, which,
when heavily armoured, as in crabs and lobsters, gives the
animal a degree of protection from its predators.

In the Malacostraca, the subclass to which belongs the order
Decapoda, which includes all the animals farmed, the animals
have bodies divided into 19 segments. The exoskeleton is also
divided into 19 segments. Between the segments there is no
deposition of calcium salts so that at this point the cuticle forms
a flexible hinge. Each segment bears a pair of appendages which
are specialized to perform a particular function. At the sides of
each segment there is an interlocking joint which means that,

unlike a fish, a shrimp can swim only by flexing the tail up and down and not from side to side.

The body can be divided into three regions; the head, the thorax and the abdomen.

The head is composed of five segments, the first two bearing the antennae and the last three the appendages used for eating. The next eight segments form the thorax, the last five of which carry pairs of walking legs, from which is derived the name of the order Decapoda, meaning ten feet. In the decapods which concern us, the head and thorax are joined to form a cephalothorax, covered by a hard carapace. The last section of the body, the abdomen, is composed of six segments and each segment carries a pair of appendages used for swimming.

To be able to grow, a crustacean has first to cast its hard outer skeleton. It grows a new soft skeleton inside its hard old skeleton which then splits along the epimeral line and the animal withdraws. It then absorbs water and swells in size, the new soft outer cuticle then forms a new hard exoskeleton, and the animal has grown.

There is no internal skeleton as such but the exoskeleton bears inwardly pointing plates and ridges to which the muscles are attached.

Muscles

The muscles of a shrimp and a lobster can be divided by their function into two types. The main body of the muscles run the length of the tail and are used to flex the tail in swimming. These muscles are the only ones of commercial importance in these animals. In the lobster and crab the muscles inside the walking legs and the powerful pincers may also be extracted and eaten.

Circulatory system

In the Decapoda the heart is a simple tube which lies along the dorsal (upper) surface of the thorax. From the heart the blood is circulated to the body by a system of arteries. There is no vein system returning the blood from the body to the heart. Instead the blood flows through a series of spaces in the body to a space surrounding the heart, called the pericardium. Blood enters the heart from the pericardium through six valves in the heart wall which allow the entrance but not the exit of the blood. As it returns to the heart, some, but not all, of the blood passes through the gills where gaseous exchange between the blood and the water occurs. This system is not as efficient as that of a fish as not all of the blood passes through the gills on each cycle. The gills are situated under the hard carapace. The gills are thus protected and are situated in a confined space through which the animal can pump water to increase the efficiency of respiration.

The respiratory pigment of the Decapoda contains copper. It is blue in colour and is called haemocyanin. The haemocyanin is not carried in the blood in special cells but is dissolved in the blood fluid.

Digestive system

The gut can be divided into three parts, the foregut, the midgut and the hindgut.

The foregut consists of two chambers with folded walls which

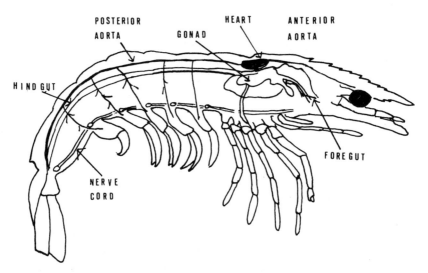

Fig. 3.3 Main internal organs of a penaeid shrimp.

carry calcified plates. The animal tears its food with the feeding appendages of the head and passes it through the mouth into the foregut, where it is ground up by the calcified plates.

The midgut is the region which contains various tubules opening from the gut wall and also a large glandular structure which is called the hepatopancreas, the function of which, although it is known that it is connected with the digestive process, is not fully understood.

The hindgut stretches from the midgut down the length of the tail to the anus; its function is also not well understood.

Excretory system

The Crustacea do not possess organs comparable to the vertebrate kidney. Excretion appears to be carried out by a variety of glandular tissues, the most important of which are situated in the head and associated with the antennae and the eating appendages.

Reproductive system

The sexes are separate. The gonads are situated in the thorax and consist of a pair of hollow organs which communicate with the exterior by ducts. The male forms the sperm into discrete lumps called spermatophores, which are attached to the female and release the sperm when she lays her eggs.

Endocrine system

The Crustacea have a well developed endocrine system. Three sets of glands have been identified in the Decapoda; the X-organ-sinus gland system, the pericardial organ system and the post commissural organs.

X-organ – sinus gland system

These glands are situated in the eye stalks and are endocrine organs of great importance to the aquaculturist. This is the major endocrine organ complex and is in many ways probably equivalent to the vertebrate pituitary. At present four hormones have been identified as being produced by this gland which, it is suggested, have about 14 different functions in the physiology of the animal. Of these, two are of importance to the farmer. These are the hormone which controls moulting and that which controls the ripening of the gonads. The existence of this latter hormone is the reason why removal of an eye stalk has been found to affect the sexual maturation of the animals, and so is used as the basis of the technology to obtain ripe animals under farm conditions.

Pericardial organs

These lie in the pericardial cavity and produce hormones, but their function is unknown.

Post commissural organs

These lie in close proximity to the nervous system. Their function is not understood although it is known that they produce one hormone concerned with the pigment of the animal.

Nervous system and sensory systems

The decapods have a well-developed nervous system consisting of a brain-like concentration of nervous tissue in the head and a ventral nerve cord which runs the length of the body.

The brain is composed of three parts and sends nerves to the feeding appendages, the eyes and the antennae. Pairs of nerves leave the nerve cord in each segment and innervate the various parts of the body.

Photoreceptors

The decapods have a pair of eyes situated at the end of the eye stalks on the head. These are unlike vertebrate eyes but are similar to the compound eyes of insects. The vertebrate eye has one lens and one retina onto which the received image is focused by the lens. The whole eye can be moved to follow moving objects. The compound eye is composed of many very small eye-like structures. The whole eye, which cannot be moved, is covered with a transparent cornea, a continuation of the body cuticle. Under the cornea are many cone-shaped organs which form the facets of a sphere-shaped eye. Each separate organ is lined with a retina and connected to the brain by a nerve. The definition, the clarity of vision, of a compound eye such as this is very difficult to determine but it is very good at perceiving movement as the image passes from one facet of the eye to the other, a fact which is obvious when one tries to catch a fly.

Chemo-thermoreceptors

Aquatic crustaceans possess well-developed chemo-thermo receptors. These organs are scattered at different points in the external cuticle of the animals, most of them being in the form of hair-like protrusions. These receptors are very sensitive and respond to chemical substances, temperature and pH (alkalinity and acidity). They are also used to locate members of the opposite sex. Some crabs can detect food extracts in the water at dilutions of 10^{-17}. The sensitivity to temperature changes is not very great and the animals can detect changes of the order of 1°C. The animals' sensitivity to changes in the chemistry of the water is important to shrimp farmers as this will affect their ability to detect food and also their reaction to unfavourable conditions.

Mechanoreceptors

The animals have mechanoreceptors present in the cuticle which respond to touch and to water movement. In the Decapoda there is an organ of balance at the base of the small antennae. This organ consists of a pit lined with sensory hairs and containing sand grains. The position of the sand grains on the sensory hairs allows the animal to be aware of its position relative to the pull of gravity. This fact can be demonstrated by giving a newly moulted animal iron filings instead of sand;

when these enter the pit the pull of a magnet causing the iron filings to rise up will cause the animal to swim upside down.

3.3   Molluscs

Skeletal and muscular system

The skeleton of bivalve molluscs is an exoskeleton as in the Crustacea, taking the form of a shell, a hard calcareous outer coating to the animal. The shell is in fact composed of three layers, an outer protein layer and two inner layers formed from calcium salts. The shell is in two halves, or valves, connected by a flexible hinge and can be opened or held tight shut by a powerful muscle system. The shell, which is formed by the special layer of tissue called the mantle, increases in size as the animal grows and can, to some extent, be replaced if it is damaged. In some bivalves the inside of the shell is covered by a smooth shiny substance called mother-of-pearl. Some species cover unwanted particles which find their way inside the shell with a layer of deposition of mother-of-pearl, and these then form what we know as pearls.

Bivalves are able to move about only very little or not at all. Some species bury themselves in the substratum and some can swim. However, most of the species of importance to aquaculture do not move after they reach adult form. Because of this there is no need for a complicated muscular system.

Circulatory system

Bivalves have a simple heart which pumps blood through arteries to the various parts of the body. The blood returns to the heart through simple sac-like veins which connect into a cavity surrounding the heart. The blood is oxygenated by passing through an extensive and complex gill system which forms a large proportion of the animal's body. This extensive system is required in order to perform the second function of the gill, which is to filter from the sea water the food for the animal.

The blood of bivalves does not contain a respiratory pigment and the oxygen concentration in the blood is the same as in the surrounding sea water. As the animal does not move this system is adequate to meet its oxygen requirements.

Digestive system

The feeding and digestive system of the bivalves is one of the most complicated systems that has evolved in any animal. The whole system is dependent on a highly organised use of cilia. Cilia occur in most animals and are microscopic hairs which line many cavities; they are, for instance, present in our lungs where they are responsible for the ejection of unwanted fluid and foreign matter. The cilia work in a thin film of liquid and as they all beat together in the same direction in waves they can drive fluids and particles along a surface. Feeding in bivalves starts in the gills, which are composed of sheets of folded series of fingerlike filaments of tissue covered in cilia.

Water, containing suspended material including the food of the animal, is pumped in large quantities through the gill complex. The suspended material is trapped on the gills by a covering of mucus and is then passed along a series of pathways lined with cilia to the mouth. At the mouth are flaps of tissue covered in cilia. The cilia systems on these flaps sort out the particles into unwanted material and food. The unwanted material is cast out into the mantle cavity and then expelled from the shell. The food material passes through the mouth and down a short oesophagus into the stomach. The stomach is a sac with finger-like outgrowths in which digestion takes place. The stomach is lined with cilia which perform complicated sorting and mixing motions of the food. The stomach contains a large organ which is called a crystalline style. This is a rod made of protein which projects into the stomach; it is continuously being secreted and is composed of protein containing digestive enzymes. The style is rotated by the beating action of cilia and, as it rotates, it rubs against a hard plate in the wall of the stomach. The rod appears to perform two functions. First, it helps to stir the stomach contents and, secondly, it releases the digestive enzymes into the cavity. A truly remarkable system!

The intestine is a short tube which conveys the waste material from the stomach out of the body.

Excretory system

Bivalves have simple kidneys which take blood from the pericardial cavity and pass it down a tube where nitrogenous waste products are excreted into it. These tubes also have the capacity to extract required salts from the fluid.

Reproductive system

The system in bivalves is very simple, consisting of paired gonads with short ducts to the exterior and no associated glands. The eggs and sperm are released directly into the sea water. The number of eggs produced by the female varies with the species, some oysters producing about 1 000 000 eggs and mussels up to 10 000 000. The eggs are fertilized by the sperm in the water. In the case of the European oyster the eggs are held inside the mantle cavity of the female and fertilized there. The eggs hatch into free swimming larvae which feed on planktonic algae. The larvae may go through two changes of form before settling on the bottom and changing into a small adult form called, in the case of oysters, a spat. This final stage settles in a natural position that the animal will keep for the rest of its life.

Some of the bivalves, including the oyster, change sex more than once during their life. The oysters retain the eggs in their mantle cavity where they are fertilized by sperm brought in by the feeding current. The oyster releases developed larvae which soon form spat, a fact which is of great value in the development of hatcheries for these animals.

Endocrine system

Very little is known about the endocrine organs of the bivalves or of any hormone produced by these animals.

Nervous and sensory systems

The nervous systems are simple but well developed, consisting of a series of nerve ganglia connected to a network of nerves.

Little is known about the sense organs of bivalves although much work has been done on the most advanced mollusc, the octopus. The bivalves are sensitive to touch, and in some light-sensitive spots like eyes are present along the edge of the mantle.

Chapter 4
Growth

4.1   Growth of fish

The growth of fish and the factors which influence growth are of paramount importance to the fish farmer, as the maximum growth of fish in the minimum time with the minimum food is his main objective.

As we shall see in the chapter on nutrition, an animal is very like an engine in that it needs fuel to function and a system to get rid of the burnt fuel. In the case of an animal this fuel is the food that it eats which it 'burns' to release energy. This energy is then used to supply the everyday functions of the fish, such as breathing, locomotion and growth. In the case of warm-blooded animals energy is also needed to maintain a constant body temperature different from that of the environment. As we have seen, however, a fish is a cold-blooded animal and it therefore does not have to expend energy in maintaining its body temperature. In this respect a fish is a more efficient converter of food to body protein than is a warm-blooded animal. However, a fish suffers a drawback by being cold-blooded. As the body temperature of a fish varies with that of its environment, and as the growth rate is affected by body temperature, then this means that a fish does not grow as fast in cold water as it does in warm water. Consequently, in many countries which have a wide seasonal variation in water temperature, fish may not grow much during the cold season. It should be pointed out, however, that the optimum water temperature for growth will vary with different species and with the ambient water temperature to which they have become acclimatized.

Food supplies the energy with which a fish grows but a fish must also use the energy of food for all its body functions.

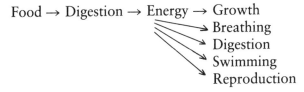

Obviously some of these functions are of a higher priority than others to the fish. Thus if energy supplies are limited, breathing and digestion are of primary importance, swimming to catch food is next, reproduction is next and last comes growth. Thus unless a fish is healthy, well fed and in a good environment its growth will be limited.

Growth also varies with body temperature. This means that with higher temperatures a fish will grow faster; however, after a certain temperature, which varies according to species, further rises result in a fall in the growth rate (Fig. 4.1). This means that in a country with seasonally changing water temperature, when the water starts to warm after the winter the growth rate starts to accelerate and continues until a certain temperature is reached, after which further rises slow down the growth rate. This cycle is then retraced as the water cools in the winter.

Accordingly, the growth rate of a fish at any given time is governed by the interplay of many factors such as food, water temperature and oxygen levels in the water. We understand

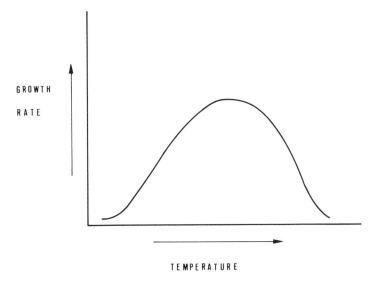

Fig. 4.1 Relationship between growth and temperature.

many of these factors and a good fish farmer must always have these in mind. We also know however that growth can be affected by factors about which we, as yet, know little.

Two of these are of great importance. The first is the effect of the density of the population of fish in a pond. We know that beyond a certain level of population density the growth rate can be adversely affected even though other factors such as food and temperature are not limiting. We do not fully understand the cause of this. The second factor is the size of the pond in which the fish are living, which we know can directly affect the growth rate of the fish and the final size that it will reach. An extreme example of this is the size that a fish will reach in a small aquarium compared with what it could attain if kept in a large pond. We do not know the mechanism of this; it is not of much significance to a fish farmer but is another example of the effect of the environment on growth.

Most animals, when they reach a certain age, stop growing, with the extent of their final size being determined by heredity. In this respect a fish differs from other vertebrates in that it never loses its capacity to grow. As a fish gets older its growth rate slows down, but a stimulus, such as a change of environment, can start a fish growing again at a faster rate.

We measure the growth rate of a fish by recording its length and weight at regular intervals, the difference over a given time being a measure of its growth. However this type of measurement does not allow us to compare the growth of two fish of different sizes as obviously a larger fish has to grow more than a smaller fish in any given time interval to grow at the same rate. To be able to compare the growth rate of different fish we use the Specific Growth Rate, which is calculated in the following way:

$$\frac{L_2 - L_1}{T_2 - T_1} \times 100$$

L_1 is the length of the fish at time T_1 expressed in natural logarithms and L_2 is the length in natural logarithms at time T_2. By using this formula we can compare the growth of any two fish, over any time period.

The growth of fish is influenced by internal environmental factors as well as by the external environmental factors discussed above. The internal factors include the hormones produced by the fish (discussed in Chapter 3) and also the genes of the fish.

As is well known the shape, size and general characteristics of an animal are determined basically by the genes inherited in equal quantities from the parents at the time of fertilization of the egg by the sperm. A normal individual is said to be diploid in that it possesses two sets of genes. Research work in fish has demonstrated that it is possible, by heat treatment of the egg, to produce an individual which has three sets of genes, called a triploid. A general characteristic of such an individual is that cell size increases owing to the extra chromosomes and results in increased body size. Work with *Sarotherodon aureus*, in which triploid individuals were produced, showed that after 14 weeks triploid fish were 33% larger than were the diploid individuals in the same brood. The development of this technology will be of future importance to the fish farmer.

4.2 Growth of crustaceans

As has been explained, increase in size occurs during the short moult period, which is the interval between casting their hard shell and the hardening of their new one. At each moult the calcareous shell is cast and so the animals do not carry any growth marks by which we can measure their age.

The growth is dependent on the frequency of moulting and the size increase achieved during the moult period. These two things are affected by various factors. Light and salinity appear to have an influence but the actual effect has not been established. Food supply and temperature are both known to be of importance. Growth can be restricted by insufficient food or food lacking in certain dietary requirements, which are responsible for many failures in shrimp farming. The development of manufactured pelleted food for shrimp has improved this situation in many farms though these foods may still be too expensive for low density artisanal-type farms.

Higher temperature shortens the intermoult periods, thus encouraging more frequent growth periods; however the effect on the actual growth achieved by the animal at each moult appears to be variable. The shortening of the intermoult period is a dominant factor and as a general rule growth increases with rising temperature within the temperature range tolerated by the species. As the shrimp gets older its rate of growth slows down considerably, but as farmed shrimp are harvested before this stage is reached it is not important to the farmer.

4.3   Growth of bivalve molluscs

As with fish and crustaceans, temperature and food supply influence the growth of molluscs. However the great problem with growth studies is the difficulty of measuring growth without killing the animal, thus making it impossible to follow the growth of individual animals. The growth of the bivalve shell can easily be determined as the shell often shows growth rings similar to those on a fish scale, but it is the growth of the flesh which is important and this cannot be measured without first killing the animal. There are very few facts established concerning the shell-to-flesh ratio and the factors which influence this ratio but obviously the food supply must be one.

4.4   Ageing of fish

Fish growth varies according to environmental conditions, season and age. When it is possible to measure fish periodically

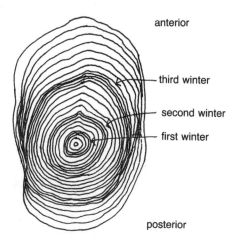

Fig. 4.2 Scale, showing features of the life cycle (schematic).

it is possible to calculate their growth. There is also a method of calculating the age of a fish and of obtaining an indication of its growth at different times of its life. This system is very like calculating the age and growth pattern of a tree by counting the rings in the wood after the tree has been cut across. In the same way a fish develops growth marks during its life and we can use these marks to determine its age and to gain some indication of its growth pattern. A fish living in temperate or semi-tropical conditions, where the water temperature varies regularly in a seasonal fashion, varies its growth rate with the changes in water temperature. These patterns are reflected in the rate of growth of various bony structures in the fish which develop rings in their structure. These structures are the opercular bones, the scales and, to some extent, the backbones of the fish, all of which can be treated to show the growth rings very clearly.

When the rings are counted a good estimation can be obtained of the age of the fish and the distance between the rings can be used to measure approximately the growth which was achieved during that period. This facility is of great value to fishery biologists who can obtain much information about age and thus the age group compositions of fish populations.

This method can be applied to the shells of some molluscs.

Chapter 5
Nutrition

5.1 Fish nutrition

The energy that a fish requires comes from its food; this energy however comes originally from the sun. The sun's energy is converted into food by plants which use the energy to produce carbohydrates. Herbivorous animals eat the plants and use the stored energy for their own use. Plants are therefore known as primary producers. Finally, carnivorous animals eat the plant-eating animals and use the energy that the first animal obtained from the plants. Thus the basic need is for energy and all the energy in living organisms comes from the sun.

It is interesting to note that man has a second form of energy, oil, and that the energy of this also originated from the sun as it was formed by little animals and plants which trapped the sun's energy many millions of years ago. When these little animals and plants died their bodies were converted to oil in the course of time.

Food is composed of two types of substances, organic compounds which contain carbon and inorganic compounds which do not.

5.2 Inorganic compounds

Minerals have many uses in a fish. The following elements, which in combination make up various minerals, are essential for its life: calcium, phosphorus, sodium, molybdenum, chlorine, magnesium, iron, selenium, iodine, manganese, copper, cobalt, and zinc.

Minerals provide strength and rigidity to the fishes' skeleton. They are essential in maintaining the correct balance of body fluids and for the correct functioning of the nervous system and the endocrine system. Iron is an essential part of the blood pigment and therefore plays a vital role in respiration.

5.3 Organic compounds

Proteins

Proteins are complex organic compounds built from basic units called amino acids. Amino acids are simple organic compounds which, when put together in different configurations, can be built to form different proteins. When the proteins are digested they are split by enzymes into their basic units of amino acids. The amino acids are absorbed through the gut wall into the blood and are resynthesized by the body to build new protein to form new tissue, for growth or to repair old tissue. Proteins can be 'burnt' for energy production.

About 23 amino acids have been isolated from animal tissue. A fish is capable of forming 10 of these for itself. The rest of the amino acids which a fish requires but cannot form itself are called essential amino acids and have to be included in the diet. These essential amino acids are required in different minimum quantities. In natural feeding, such as a fish gets from an adequate supply of natural food, this presents no problems. When, however, a fish is fed on artificial food careful formulation is required to ensure that these dietary requirements are met.

Fats

Fats are composed of basic units called fatty acids. Some fats are digested from the gut lumen by being emulsified into micro-droplets with the aid of bile secreted by the liver. Other fats are broken down by the digestive enzymes into their constitutent fatty acids and then absorbed by the gut.

Fat is the main form of stored energy in an animal. When food is abundant fat is laid down to be used later during food shortages and during periods of extensive exercise. Detailed information on the fat requirement of fish is lacking, but it is known that certain fatty acids are essential for health and growth and these have to be provided in artificial diets.

Carbohydrates

Carbohydrates are a broad group of substances comprising sugars, starches and celluloses. The simplest carbohydrates are sugars and the most complicated are polysaccharides. Polysaccharides are of two sorts, so-called structural poly-saccharides, cellulose and chitin and digestible polysaccharides such as starch.

The ability of an animal to digest starch depends on whether or not it produces the enzyme amylase. This enzyme is found mainly in those fish which eat a diet of plant material. The principal product of carbohydrate digestion is glucose, a very important substance as it forms the basic fuel for the body tissues.

Vitamins

Vitamins are organic compounds essential to various aspects of the health of an animal. They can be divided into two types, water soluble and fat soluble.

Water soluble vitamins

Thiamine, found in cereals, legumes, yeast and animal tissue – deficiency causes poor growth and convulsions.

Riboflavine, widely distributed in animal and plant tissue – deficiency causes poor growth, eye complaints and dark coloration.

Pyridoxine, found in yeast, liver and cereals – deficiency causes anaemia, nervous disorders and respiratory problems.

Pantothenic acid, found in animal and plant tissue – deficiency causes poor growth, clubbed gills and necrosis.

Inositol, found in animal and plant tissue – deficiency causes poor growth and skin lesions.

Biotin, found in animal tissue – deficiency causes poor growth and muscle atrophy.

Folic acid, found in animal and plant tissue – deficiency causes poor growth, dark colour and anaemia.

Choline, found in animal and plant tissue – deficiency causes poor growth and haemorrhagic conditions of the kidney and intestine.

Nicotinic acid, found in animal and plant tissue – deficiency causes poor growth, weakness and loss of appetite.

Vitamin B12, found in animal tissue – deficiency causes anaemia.

Ascorbic acid, found in fruit and vegetables and in animal tissue – deficiency causes eye lesions and haemorrhagic lesions in the tissue.

Fat soluble vitamins

Vitamins A, D, K and E.

Water soluble vitamins are taken in by an animal, metabolised and then excreted but fat soluble vitamins are taken in by the animal and stored, not excreted; thus an excess of these vitamins in the diet can cause problems.

Vitamin A, found in fish oils – deficiency causes poor growth and poor vision.

Vitamin D, formed by animals in their skins by the action of ultra-violet light; it is found in animal tissue and in fish oil – deficiency causes abnormal bone formation.

Vitamin E, found in vegetable oils – deficiency causes anaemia and poor growth.

Vitamin K, found in vegetable tissue – deficiency causes impairment of the blood-clotting mechanism.

5.4 Factors which influence energy requirements

Temperature

As a fish is cold-blooded the rate of functioning of its body quickens with mounting temperature. Therefore, with higher temperatures more energy is required.

Above a certain temperature, which differs according to species, the fishes' metabolic processes cannot keep up with the more demanding energy requirements indefinitely. Thus, above a certain temperature, energy supplies in the fish become limiting and, apart from anything else, the growth rate of the fish falls. This effect results in there being an optimum temperature for growth.

Water flow

A fish has to expend more energy in swimming if it is kept in fast-running water than it does when it lives in a pond.

Body size

Unit weight compared with unit weight, the metabolic rate of a small fish is higher than that of a large fish, and therefore, in farming practice, it is found that small fish require more food per unit of body weight.

Feeding

The metabolism increases during feeding activity. This is especially so in intensive farming.

Other factors

A prolonged increase in physical or metabolic activity raises the energy requirement of a fish. Such increases are mainly caused by what are known as stress factors. Crowding, low oxygen levels, pollution and high ammonia levels are all stress factors met in farming which increase the energy requirement and can adversely affect growth rate.

Energy uses

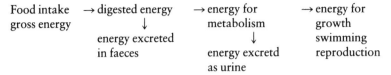

Energy obtained from food is used in the body to perform different functions and the energy requirement of some of these takes priority over that of others. Digestion, excretion, respiration and the basic metabolism of the body tissues have first call on the energy supply. Physical activity for swimming, for feeding and for escape from predators is also high in the priority list. If the total energy intake does not exceed the amount needed for these basic functions then the fish will not become sexually mature nor will it grow. Under some special conditions a fish may mature its gonads by using stored energy, thus losing weight. Fish which perform long spawning migrations without eating, such as the Atlantic salmon, are examples of this type of mechanism.

5.5   Crustacean nutrition

Our knowledge of the nutritional requirements of some crustaceans is limited as very little work has as yet been carried out in this field. We know a little about the kind of fats, carbohydrates and proteins they need but we do not know how much they require of the various dietary components. However, much work has recently been carried out on the nutritional requirements of shrimp and optimum proprietary feedstuffs are now widely available.

5.6   Bivalve nutrition

Farmed bivalve molluscs are herbivores which obtain their nutrients by filtering microscopic plants (phytoplankton) out of the sea water in which they are kept. Bivalve farming, unlike farming of carnivorous fish and crustaceans has the advantage that the adult animals do not need to be supplied with additional food. However, larvae and seed produced in hatcheries (see Chapters 23, 24, 25) are fed on specially reared micro-organisms.

Chapter 6
Health of Farmed Aquatic Animals

6.1 Fish diseases

The diseases from which aquatic animals in farms suffer are many and varied and are described in specialized books. The purpose of this chapter is to give a general introduction to these diseases; also, and more important to practical farming, to outline the precautions necessary for the prevention of these diseases. The diseases of fish in farms can, in the main, be treated but treatment is often of doubtful value. The cost of the treatment can be high. Sick fish do not grow and so the farmer loses money as marketing is delayed. Also, sick fish die; this is of course then a total loss to the farmer, not only the loss of the fish and its anticipated value but also the loss of the money invested in the fish, both capital costs and running costs. If the fish is near market size when it dies, these losses are severe. It is therefore much better to prevent disease; prevention is cheaper than disease treatment and avoids loss due to poor growth and death of sick fish.

Diseases can be classified under the following headings: viral, bacterial, ectoparasitic, endoparasitic, fungal and nutritional. Viral, fungal, bacterial and parasitic diseases are all water-borne and can be carried from pond to pond by the introduction of new fish or by the farmer and his equipment. Nutritional diseases are caused by shortages of essential components in the diet. Parasitic diseases are caused by animals which often complete their life cycle in animals other than fish or in different fish species.

The following general review of some of the more common diseases encountered in aquaculture will give an appreciation of the problems and will form an introduction to the specialised text books.

Viral diseases

Little is known of viral diseases in wild populations; they are however of great importance in farming conditions.

(1)   Infectious dropsy causes inflammation of the abdominal organs and the accumulation of liquid in the abdomen. It can be treated with tetracyclin, or chloramphenicol or nitrofuran.

(2)   Viral haemorrhagic septicaemia (VHS) is a disease of trout which causes exophthalmia, anaemia and a swollen abdomen.

(3)   Lymphocystis causes an enlargement of the cells of the connective tissue and a consequent growth of this tissue in the skin and fins. This disease has a world-wide occurrence in both freshwater and marine species.

(4)   Infectious pancreatic necrosis (IPN) occurs in trout where it causes necrosis of the pancreas; it is highly contagious and is fatal.

(5)   Fish pox is caused by a virus of the herpes group and causes papillary growths and lesions in the skin and eyes.

Bacterial diseases

(1)   *Aeromonas liquefaciens* causes haemorrhagic septicaemia and can be treated by using sulpha drugs or antibiotics.

(2)   *A. salmonicida* causes furunculosis in salmonids. This disease is of considerable economic importance and is treated with sulpha drugs or the antibiotics chloromycetin, terramycin or nitrofuran derivatives.

(3)   *Pseudomonas* infections are widespread – for example 'spot disease' in carp. Treatment is by terramycin or chloramphenicol.

(4)   *Vibrio anguillarum* causes skin lesions in most salt water fish. Treatment is by nitrofurans.

(5)   *Chondrococcus columnaris* causes skin lesions. This is an important disease in aquaculture, as fish are more prone to it if they are living in poor environmental conditions. Some success has been achieved in treatment with sulpha drugs; work is also being carried out on the development of an oral vaccine.

(6) Myxobacteria cause gill disease, resulting in the fusion of the gill lamellae. Treatment is with the antiseptic pyridyl mercuric acetate (PMA).
(7) Corynebacteria cause kidney disease in salmonids. The disease can be controlled by the use of sulpha drugs or antibiotics.
(8) *Haemophilus piscium* causes skin ulcers in salmonids and Indian carp. These infections have been treated with sulpha drugs or tetracyclines, also by the sterilization of the water with chlorine.
(9) A form of tuberculosis occurs in fish which causes wasting of the body and is treated with kannamycin.

Diseases caused by parasites

It is very difficult to estimate the economic importance of this type of infection. All parasites occur in wild fish but the level of infestation can become high under farm conditions and so significantly affect the performance of the fish as a farm animal.

Protozoans

(1) *Ichthyophthirius multifiliis* causes ichthyophthiriasis in trout; this takes the form of irritant white spots on the skin. This disease is of major economic importance in trout farming, where it causes considerable damage. The parasite has a simple life cycle which can be broken by the use of salt or other chemicals.
(2) *Cryptokaryon irritans* is the marine equivalent of Ichthyophthirius.
(3) *Trichodina megamicronucleata* causes a disease in carp. Treatment is by the addition of chemicals to the water.
(4) *Chilodonella cyprini* causes a skin disease in carp by feeding on the cells of the skin. It is treated by increasing the salinity of the pond water.
(5) *Hexamita salmonis* causes a disease in salmon by damaging the cells of the caecal mucosa. It is treated by the addition of chemicals to the pond water.
(6) *Costia necatrix* attacks the skin of many different host species. It is eradicated by disinfection of the farm ponds.

(7) *Oodnium ocellatum* attacks fish living in farm conditions. It is treated by the use of low concentrations of copper sulphate in the pond water.
(8) *Myxosoma cerebralis* causes 'whirling disease' in salmonids. It infests the semi-circular canals in the ear of the fish. The parasite produces highly resistant spores which can live for many years in the pond bottom. The disease can be combated by treating the pond bottom chemically to break the life cycle.
(9) *Myxobolis cyprini* causes cysts in the skin and body of carp; treatment of this disease is difficult.

Trematodes

Monogenea   These trematodes have only one host species. The wounds which they cause lead to secondary fungal infections and they are responsible for heavy fish losses in carp farms. There are two types, dactylogyrids, which attack the gills and skin, and gyrodactylids, which attack the gills. They can be controlled by the addition of chemicals such as formalin and potassium permanganate to the water.
Digenea   These trematodes have several hosts.

(1) *Sanguinicola inermis* is a minute fluke which inhabits the blood vessels of fish and causes serious damage.
(2) *Posthodiplostomum cuticola* causes black spot disease in carp. The fish is the intermediate host, a snail the primary host and a heron the final host.
(3) *Crepidostomum farionis* is an intestinal fluke in trout which can be treated with di-N-butyl tin oxide given in capsules which dissolve in the fish's stomach.

Cestodes

Tapeworms are not apparently of great economic importance in aquaculture although certain species can cause problems. They can be controlled by interrupting the life cycle or by the use of anthelmintics such as di-N-butyl tin oxide.

Nematodes

These occur widely in fish but are of no great economic importance in aquaculture.

Leeches

These do not cause serious problems in aquaculture.

Crustaceans

(1) *Lernaea cyprinacea* causes heavy losses in catfish farms. Treatment is by the addition of chemicals to the water, including salt, formalin and organophosphate insecticides. Lernaeids have world-wide distribution and cause problems in aquaculture. In Japanese eel farming, for example, they attack the mouths of the fish.
(2) *Argulus sp.*, the fish louse, attacks the skin and causes problems of secondary infection.

Fungal diseases

Examples include *Saprolegnia*, which attacks fish skin, and *Ichthyophonus*, which occurs internally. Fungi can be treated with fungicides, especially malachite green, in the water. In fresh water, they can be treated with salt.

Algae

Algae are of no direct significance but they can cause heavy mortality in fish ponds when outbreaks of such species as *Prymnesium parvum* occur as these produce toxic substances. The algae can be killed by the use of copper sulphate but the dead algae again cause problems by reducing the oxygen levels in the pond.

6.2   Preventing fish diseases

There are certain basic rules which must be observed if outbreaks of disease are to be prevented or, if they occur, contained.

The water supply to each pond must be separate. It is very
bad farming practice to supply a pond with water from
another. Water from a fish pond may carry disease and will
probably be deficient in oxygen, and with a high level of
metabolic products such as ammonia.

Fish must not be stressed. When they are handled, unless
they are due for market, the greatest care must be taken to
upset them as little as possible. This is necessary for two
reasons. First, it is well established that extreme stress can be
the direct cause of fish death even up to a week or more after
the stressing occurred. Secondly, harsh handling of fish
damages their skin, rubbing off the scales and the protective
slime. This can allow the entrance of disease organisms.

Fish must be kept in optimum conditions at all times. They
must be kept in water with plenty of oxygen, with the correct
pH, and with a low ammonia content. The optimum water
quality requirements beyond which the health and growth of
the fish with be adversely affected varies considerably from
species to species. An indication of the range of these
requirements is given in the following table.

	Tilapia	Carp	Salmonids
Temperature range	8–40	16–40	below 23
Lethal oxygen limit (mg/litre)	2–3	3	5
pH tolerance	5–11	5–12	5–10
Lethal carbon dioxide concentration (mg/litre)	73	—	5
Lethal ammonia concentration (mg/litre)	4 (pH 7.4)	10–13	2 (pH 8.0)
Turbidity (ppm)	13,000	190	—
Salinity (‰)		12	15

The fish must have assured food, either supplied as
manufactured feed or by natural production in the pond. The
feed must supply all the essential dietary requirements of the
fish, as outlined in Chapter 5.

Great care must be exercised when mixing populations of

fish from different ponds, or when introducing new fish into the farm so that no diseased fish are introduced. Fish new to the farm site should be held in quarantine, until it is certain that they have no disease, before being released into the farm ponds.

Separate equipment should be kept for each pond as disease is easily spread from pond to pond. All major equipment such as nets and aerators should be sterilized after use as a matter of routine. A useful agent for this is benzalkonium chloride solution containing 600 ppm of active ingredient.

Good farm practice and regular inspection of stock is essential to good farming. Wherever possible fish seed should be obtained from suppliers who can show evidence of being able to supply seed from disease-free stock.

6.3   Containing fish diseases

Large-scale intensive farming is already suffering severe problems with fish diseases. These are probably going to intensify in the future. With all the economically important diseases where drugs are used as routine measures, either as a treatment or as a prophylactic, the main danger is the development of drug resistance. The following characteristics are important for therapeutic drugs: the difference in the lethal dose for the disease organism and for the fish must be at least 1:4; the drug must dissolve in water; the price must be economic and the use of the drug must not restrict the productivity of the pond.

During the last 10–20 years considerable progress has been made in understanding the immune systems of fish. At the present time successful results have been achieved in the vaccination of salmonid fish against bacterial diseases (vibriosis, furunculosis and enteric redmouth disease). Much less is known about protection of the fish from viruses and of the protection of freshwater tropical and semi-tropical fish.

Vaccination of rainbow trout against infections of *Vibrio anguillarum* has been successfully administered orally, by injection, by high pressure spray and by immersion. The latter technique appears to be the least stressful to the fish. The fish are allowed to swim in a vaccine solution for about 2 hours. Protection was shown to last for about 5–7 months and the fish showed a reduced rate of infection during their life in the sea. The cost of the vaccine is about 2 US cents per fish of 50–75 g.

6.4   Shrimp diseases

Crustaceans suffer from diseases caused by bacteria, viruses, fungi, ectoparasites and endoparasites. As in fish farming the prevention of disease is better than any possible cure. Prevention of disease requires the maintenance of optimum pond conditions and the avoidance of injury of the animals. The same basic rules which were given for fish farming apply equally to shrimp farming.

Diseases are treated when met under hatchery conditions especially during the post larval stage, but here again prevention is better than cure and often regular prophylactic treatments are given. These treatments involve the use of chemicals and antibiotics and the use of several vaccines is now possible which will minimise the impact of bacterial disease affecting farmed shrimp species.

6.5   Mollusc diseases

Cultivated molluscs are known to be affected by diseases caused by viruses, bacteria, fungi and protozoans as well as by larger parasites and pests. For example, *Marteilia* and *Bonamia* protozoans have caused heavy mortalities in flat oysters (*Ostrea*). *Crassostroa angulata* has suffered losses from gill disease of uncertain origin. Viruses and bacteria can cause losses in hatcheries.

Spread of pests and diseases is usually prevented by restrictions on the movement of stock from affected areas.

Part 2
The Husbandry of
Farmed Aquatic Animals

Having provided the reader with sufficient background information on the biological functioning of farmed aquatic
animals, we now turn to the general methods which may be
adopted in the successful pursuit of fish farming. After de-
scribing the main farming systems in use today, we consider the
various culture practices, i.e. the ways in which fish farming has
developed as food production systems in different parts of the
world. The rest of Part 2 then works sequentially through the
various practical stages of fish husbandry, i.e. from the construction of ponds to obtaining the seed, feeding and rearing
the fish and finally to harvesting them. Although it is beyond
the scope of this manual to cover some of the wider aspects of
fish farming, e.g. predation, legal implications, economics,
marketing, fisheries extension services, etc., these are considera-
tions which are relevant before embarking upon production.

There are wide variations in husbandry techniques, not only
from country to country but also between levels of development
and technical sophistication. Every fish farm is different and the
aquaculturist usually develops his own practices which may be
constantly refined with experience.

Chapter 7
Aquatic Farming Systems

7.1 Farming systems

Farm systems are usually classified according to the density of stocking and the level of supplementary feeding. The five main types are extensive, semi-intensive, integrated, intensive and recirculatory.

In the extensive system the animals live and feed naturally in large enclosures at low densities and inputs to the system of labour, capital and other supplies are limited. Examples of this type of farming are artisanal farming of fish in rice fields and the ranching of salmon in the sea. The semi-intensive system uses large ponds which produce some natural food. This food is supplemented by the addition of feed and/or fertilizers. Almost all earth ponds fall into this category, in which densities higher than those attained in extensive farming can be achieved.

In an integrated system fish production is combined with the farming of other animals, the excreta from which fertilize the fish pond to stimulate fish food production. Supplementary food is sometimes given. Higher densities of fish can be achieved than are possible in the semi-intensive systems. In the intensive system high densities of fish are kept in enclosures and all the fish feed is given by the farmer.

The recirculatory system represents the extreme of intensive farming. In it the established ideal that water is used only once in a farming system is disregarded. The water is used over and over again, but it is treated and purified by chemicals to maintain water quality and to eradicate disease. A system using recycled water offers numerous advantages to the farmer: a controlled environment for the fish, greater freedom of site selection, freedom from water supply constraints, control over disease, and the ability to control water temperature and day length in order to manipulate growth rates and times of reproduction of the fish. In this system fish farming approaches more closely the type of control exercised by commercial poultry farmers.

However, it is expensive to install and maintain and the economics of the system limit its use to the rearing of the early stages of high-priced fish. The use of this system could be of interest in salmon smolt production.

7.2   Farming enclosures

Fish enclosures

There are many different methods of enclosing fish on a farm; earth ponds, concrete-lined ponds, concrete race-ways, concrete tanks, fibreglass tanks, netting enclosures and floating cages.

Earth ponds are the basic unit of fish farming in worldwide use and are dealt with in greater detail in other chapters in the manual. It is the essential unit in a farming system which is dependent on the natural production of fish feed in the pond, as the earth bottom forms an integral part of the ecosystem of the pond. The bottom functions in this respect in a similar way as a field in land farming. The earth pond is used in all sizes ranging from individual ponds for the production of food for the farmer and his family up to many hectares of ponds in commercial farms (Figs 7.1 and 7.2).

Concrete-lined ponds are used in systems where the whole of the food for the fish is placed in the pond by the farmer as the concrete bottom forms no part of an ecosystem. Concrete ponds are expensive to construct and are used for the intensive production of high-priced fish such as trout (Figs 7.3 and 7.4). Because concrete can be formed to make different-shaped tanks they are often specialized in design in order to support very high densities of fish. Two variations of this type are the circular tank and the raceway.

In the circular tank the water enters through a jet at the side and leaves through a central exit. The body of water is kept in rotation, causing the fish to swim continuously into the current. Because the water is rapidly and continuously exchanged, high levels of oxygen and low levels of metabolic products such as ammonia are maintained. It is therefore possible to keep very high densities of fish in such a tank. Fibreglass tanks (Fig. 7.5) are similar in design to concrete tanks. They are more chemically inert than concrete tanks and are more easily mass-produced and handled. Their availability in developing countries is, however, more restricted.

Fig. 7.1 The large scale use of earth fish ponds in Israel.

Fig. 7.2 A small artisanal farm using earth fish ponds in Colombia.

Fig. 7.3 Concrete tanks in a trout farm.

Fig. 7.4 Concrete tanks in a small artisanal trout farm.

Fig. 7.5 Circular fibreglass tanks in a commercial trout farm.

The race-way is a simpler form of tank which is long and relatively narrow. A high flow of water enters at the top of the race-way and exits at the other end. Because of this high water flow greater stocking densities can be maintained. These types suffer one great disadvantage: they are almost totally dependent on the continuous flow of water. If the water supply fails for any length of time the fish in the pond are killed. Emergencies can be met by emergency aeration but this remedy is generally only of limited duration.

Floating cages are a more recent development. They basically consist of a floating collar from which is suspended a netting bag. Water can freely pass through the netting, thus it is constantly changing. The fish are held inside at a comparatively high density and circulation of the water is enhanced by the swimming action of the fish. Cages may be rectangular, polyhedral or circular in shape and the floating collars often have a walkway around them (Fig. 7.6). Many cages may be joined to form a large complex which sometimes has a central raft carrying service equipment. The cages may be moored to the sea bed or attached via walkways to the shore. Cages sometimes have a top net covering to reduce predation. Fouling

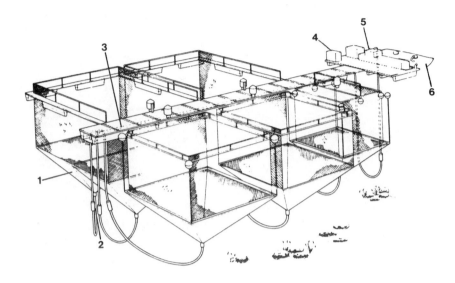

1   PVC CANVAS FUNNEL

2   AIR-LIFT PUMP

3   PIPE FOR TRANSPORTING WASTES/WATER TO FILTRATION UNIT

4   COMPUTER FOR CONTROLLING FEEDING, PUMPING, AND FILTERING
    OPERATIONS

5   FILTER UNIT

6   WASTE RECEIVING TANK

Fig. 7.6 Sea cages with controlled waste collection system (from Beveridge: *Cage Aquaculture*).

of the netting by plant matter can be a problem so cleaning and repair work is necessary. There is the additional problem of excessive excreta concentrations under the cages but this can be alleviated by remooring them in a nearby suitable location.

The main advantages of floating cages are that they provide a high volume of constantly changing water without incurring pumping or excessive construction costs. Experiments are now being made on cages which are completely submersible. They are polyhedral in shape and are better able to withstand adverse weather and/or water conditions.

Cages are the enclosures used in the modern salmon farming industry. Salmon have to live in sea water and therefore if they could not be kept in cages floating in sheltered coastal regions the alternative would be very expensive sea water pumping operations. Fish such as tilapia can be kept in cages, in rivers and lakes. However it has to be remembered that, as very little natural food reaches the fish living in cages, they have to be fed by other means. The economics of cage culture are therefore very different from those of earth pond culture of the same species.

Netting enclosures are a combination of pond and cage culture in that they are large ponds formed by a netting wall which cuts off a large area of sea or freshwater lake. The enclosure has the advantage over the cage in that natural food production can occur because the system includes the bottom of the water body. However the weakness of the system is in the netting wall which gets fouled by the growth of plants and animals; these cause its resistance to water pressure to increase at the same time as the net tends to rot; the result is a greater danger of failure of the netting barrier with a consequent escape of the fish. Although difficult, and often requiring the work of divers, great care has to be taken in continuous maintenance of the structure. Such systems are, however, in use commercially

Fig. 7.7 A floating platform attached to the shore used for mussel growing in Venezuela.

Fig. 7.8 Rafts used for mussel growing in Venezuela.

Fig. 7.9 Mussels growing attached to a rope suspended from a raft.

in eastern Asia, particularly in Japan, and therefore must be regarded as commercially viable.

Shrimp enclosures

On an extensive scale, only earth ponds are used in shrimp farming. By 1991 the area of earth ponds devoted to shrimp production throughout the world rose to over one million hectares, the four largest producers being China, Indonesia, Thailand and Ecuador. Trials carried out in concrete ponds have served to underline the usefulness of the production of natural feed in shrimp farming and this is, of course, unavailable in concrete ponds. However, now that complete shrimp diets are becoming more widely available, shrimp are increasingly being reared in various tanks though, on a world scale, the volume of this output is still very low.

Bivalve mollusc enclosures

Bivalve molluscs are not produced in farm enclosures. They cannot move about and, as they are dependent on access to unlimited supplies of plankton-rich sea water, they are kept on different types of support structures in coastal waters. At the experimental level oysters have been kept in concrete tanks and fed on cultured algae. However this technology is a long way from being a commercial process.

The objective of the bivalve farming system is to keep the animals in clean plankton-rich sea water away from sediments, which foul the feeding mechanisms of the animals, and away from predators. For this reason the majority of the world's bivalves are farmed on platforms, cages or ropes suspended from platforms or rafts, though some are laid on the sea-bed.

The raft system can be used only in sheltered sea conditions where there is an adequate depth of water at low tide. The shellfish are attached to ropes which are suspended from the rafts. In other systems the animals are attached to ropes which are suspended between posts spaced along the seashore. In yet another system the animals are kept on racks which are attached to posts along the shore. These systems are dealt with more fully in Part 3 of this manual.

Chapter 8
Fish Culture Practices

Although there are two main forms of fish farming – polyculture in which more than one species is farmed in the same
system and monoculture, in which only one species is farmed –
an additional form of culture is included here, that of 'ranching'.
In this system fish are reared in a normal aquaculture environment, and then released to lakes, rivers, the open seas or
oceans when their survival can best be ensured.

8.1 Polyculture

In an ecosystem such as a fish pond there exist various positions which can be filled by different species. These positions
are called niches and they may be due to differences of physical
position (e.g. bottom-living or mid-water-living fish) or differences between species of food preference or method of
feeding (e.g. detritus eaters or plankton eaters). For example,
mullet live mostly in the lower regions of the pond and feed
on mud and detritus which they find on the bottom. Tilapia,
on the other hand, live in the water body but some species
feed on plants, some directly on plankton, whilst others graze
off the sides and bottom of the pond and off water plants. By
combining different species in the same pond, the total fish
production can be raised to a higher level than would be
possible with only one species. With one species the availability
of its food would be the limiting factor determining the level of
productivity, but with two species in the same pond eating
different foods higher levels of fish production are possible
with the same level of fertilization.

The term polyculture can also be used to describe the integrated farming of fish with other animals. This system is con-

fined to the use of earth ponds where natural food production can be maintained. In this system animals such as pigs and poultry are kept in conjunction with a fish pond. Ducks are allowed to swim on the water and directly manure it; pigs and hens are housed next to the ponds and the manure is added by the farmer to the water. The object of these systems is to manure the ponds directly, or as near directly as possible, thus saving the haulage and labour costs of carrying manure to the farm, and also to ensure that the supply of manure is under the control of the farmer. The system of course also produces a valuable second crop for the farmer.

8.2 Monoculture

In monoculture only one species of fish is kept in the pond. This is used where very high levels of production are aimed at with the use of supplementary feeds, or, as in the case of trout and catfish farming, where all the feed is added to the pond. In such a system there is no value in polyculture as there is no food web in the pond for the different species to exploit.

There is one special example of monoculture which actually involves the use of two species of fish. This is one of the systems used to control the problem of superpopulation of tilapia ponds due to the prolific breeding of these species. One way of controlling this problem is to farm the tilapia with another species of fish which eats the young tilapia but, as it is a small fish, cannot eat the adult tilapia. In this way the size of the tilapia population is controlled.

Both monoculture and polyculture are practised in shrimp farming. In monoculture only shrimp are farmed and active steps are taken to eradicate other species which may compete for food with the shrimp or prey on them. In polyculture the shrimp are grown in the same ponds as fish, the fish used in many cases being the milkfish. The milkfish does not eat shrimp nor does it compete directly for food. The milkfish does, however, reduce the level of shrimp production, and since it is of a lower commercial value than shrimp there is no great commercial interest in this system. Nevertheless, the shrimp can form a useful cash crop for an artisanal farmer producing milkfish for his own or community consumption.

8.3   Ranching

Strictly speaking, the term 'aquaculture' should be restricted in use to describe systems in which the animal is under the control of the farmer throughout its life. However, there is one important branch of fish farming which does not fall within such a definition. This is the system where the fish are reared on the farm to a certain age and are then released without control or added food into large bodies of natural water, lakes and oceans, where they complete their life cycle and where they achieve the major part of their body growth. Such a system has one major advantage and one major disadvantage. The major advantage is that during the phase when the fish are forming their body weight, which in a farm is the most expensive in terms of capital equipment, ponds etc. and food, the fish are feeding in the wild at no cost to the farmer. The major disadvantage is that, having been released, the fish have to be recaptured for harvest.

The simplest form of this type of system is when the fish are released into a lake, where they are captured by an inland water fishery. A good example of this type of development can be seen in Brazil where major hatcheries have been constructed in conjunction with large hydro-electric reservoirs. These hydro-electric schemes cut the rivers and thus prevent the migration of the stocks of fish during their spawning run. The river stocks are therefore maintained artificially by the hatchery production. Lake stocking systems of this type, if they are to develop and maintain a successful fishery, have to be accompanied by stock assessment programmes to ensure that over-population of the lakes does not result.

A different type of stocking programme has been developed, particularly in Scandinavia and in the USA, where hydro-electric schemes prevent the ascent of the breeding populations of salmon up the rivers. Large hatcheries have been built which are used to rear young salmon from the eggs taken from the ascending fish. The young salmon are released into the river to return to the sea to complete their growth phase. When they become mature these fish in turn return to the river of origin to breed, which is the habit of all adult salmon.

This facility of the salmon to return to its river of origin has been exploited in another variation of salmon ranching. In this system, which was developed in the north-west of the USA and has now been developed in New Zealand, Chile and Japan, salmon are reared in large numbers to be released into the river. This is not a re-stocking programme but is solely to develop a fishery of the returning mature salmon which have grown at no expense to the farmer. This is a particularly advantageous system for use with the Pacific salmon species, which have a very short period between hatching and being capable of entering the sea. The one obvious disadvantage is that the farmer has no control over his valuable adult stock. Many fish are lost at sea, particularly if commercial fishermen discover the feeding grounds of these fish, as has been the case with the Atlantic salmon. However the economics of the system appear to be attractive. It is of particular interest that this form of ranching is being developed in the southern oceans as these seas appear to have no fish occupying the same niche as the salmon. The southern oceans, which are particularly rich in plankton, could possibly support a very large feeding stock of salmon.

In Japan advantage has been taken of the fact that the fertile inland sea is very nearly an enclosed area of water. Into this water the Japanese release shrimp for ranching and also young of the scallop, the swimming bivalve. Such ranching development is presently possible only in this very special region though experiments are being carried out elsewhere in sheltered marine waters (for example, the Danes are currently liberating young cod into parts of the Baltic Sea).

Chapter 9
Farming in Earth Ponds

Attention here is concentrated on earth pond construction, rather than the construction of more sophisticated aquacultural systems because, world-wide, these systems are easily the most numerous and are generally the easiest to construct. They are also the cheapest per unit area of pond and are the most adaptable in terms of shape, size and method of construction. Earth ponds are likely to remain the dominant production system in most developing countries, though in the longer term it is likely that there will be a trend towards more artificial production environments.

9.1   The construction of earth ponds

Siting

The ideal position for a pond is one where it can receive a water supply under gravity and discharge the used water under gravity. Such a situation is shown in Fig. 9.1.

The water is fed to the ponds from a small dam constructed across the stream upstream from the site. After passing through the pond the water is discharged to waste downstream.

Soils

Obviously impermeable soil is the best with which to construct a pond. For this reason clay soils are preferred. With relatively light soils it is often possible to mix clay to give a watertight covering to the pond bottom. Sandy soils and rocky soils are unsuitable. Sites in which there are tree roots or roots of plants

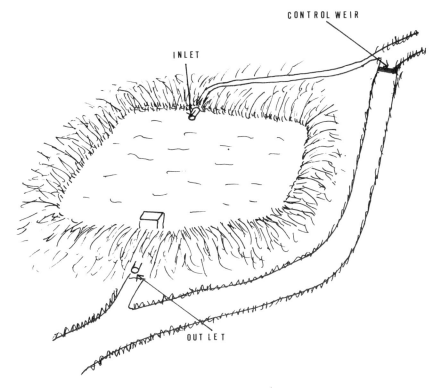

Fig. 9.1 Siting a pond.

such as reeds can give problems as, when the roots rot, they form escape holes for the water.

Pond construction

As a general rule, a pond is constructed by excavating the pond area and using the excavated material to form the pond dykes. Depending on the topography of the site, these dykes may have to be constructed on one to four sides of the pond. The correct construction and the slope of the dykes is the key to the successful establishment of a pond.

Dykes

The dykes are built up of compacted layers of soil and each should be spread not more than 20 cm thick before compaction.

Fig. 9.2 General view of an earth pond under construction. The pond walls have yet to be constructed and compacted.

Fig. 9.3 The use of a bulldozer to level the site of the future earth pond. Suitable material is dumped at the edge of the site for use in the pond wall construction.

Each layer must be carefully compacted before the next is laid. Dykes must never be built from heaps of material pushed into position and compacted. When dykes are being built on permeable soil they have to be constructed of impermeable soil or with a central core of impermeable material some 50 cm

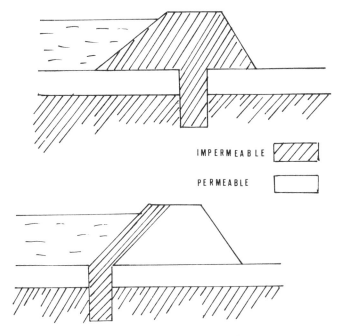

IMPERMEABLE

PERMEABLE

Fig. 9.4 Cross-sections of two different dyke constructions.

thick. This core must go down below the level of the pond bottom to form an impermeable barrier. Alternatively a dyke constructed of permeable soil can be faced on the inside with impermeable material. In this case the impermeable layer must be carried down below the pond bottom.

Slope

The slope of the sides of the dykes is very important as it determines the resistance of the dyke to erosion by rain and waves. The width of the top of the dyke should never be less than 1 m but it is often wider to allow for the construction of an access road along the top (Fig. 9.4).

The slopes of the inside of the dyke should be 1:2 but can be decreased up to 1:4 in large ponds where wave action will be greater or if light soil is used in the construction. The slope of the outside of the pond is 1:1 or 1:1.5 (Fig. 9.5).

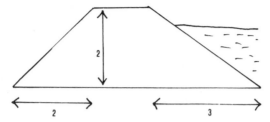

Fig. 9.5 Relationship of dyke dimensions, forming slopes.

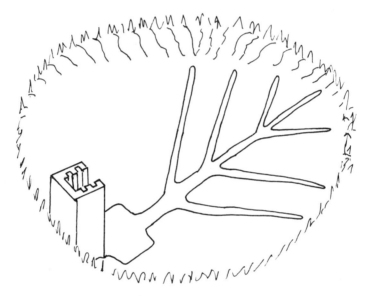

Fig. 9.6 Pond bottom drainage system.

Pond bottom

After construction is completed the pond bottom is smoothed and compacted and then a system of drainage ditches is dug into it (Fig. 9.6). The principal ditch runs down the length of the pond ending at the outlet. The bottom of this ditch is about 50 cm wide and the sides slope about 1 : 1.5. The minimum fall along the length of the ditch is 1 per 1000 but falls of 2 or 3 per 1000 give better drainage. The central ditch drains the sides of the pond bottom via a series of herring-bone ditches spreading out from the central ditch to the sides of the pond. The fall of

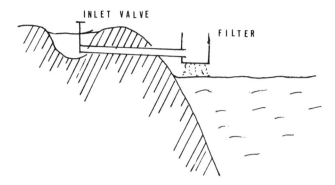

Fig. 9.7 Inlet to a pond.

these secondary ditches is 5 per 1000. In front of the pond outfall is constructed a collection basin sunk below the level of the pond bottom and lined with concrete, thus forming a firm base for the collection of fish.

Inlets

Inlets to ponds are generally simple structures which enable the water supply to be turned on and off and which prevent the entrance to the pond of unwanted organisms. This entrance is prevented by passing the incoming water through a trap of fine material which filters out the eggs and other stages of undesirable species (Fig. 9.7). It is most important that the inlet filter is regularly checked and cleared so as to prevent blockage of the water flow.

Outlets

The outlet of the pond should be formed by a structure which is called a monk (Fig. 9.8).

The monk is generally constructed of concrete although it can be made of wood or bricks. The outside edge is connected to a pipe which runs straight through the dyke to the outside. The sides of the monk on the inside of the structure have vertical grooves in which are placed wooden blocks to form the fourth

Fig. 9.8 A monk outlet.

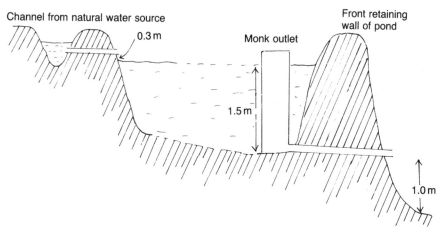

Fig. 9.9 Schematic cross-section along length of a pond.

side of the box. The water level in the pond is controlled by the height of the wooden blocks. The space between the blocks is often packed with clay or sawdust to make the wall watertight. When the pond is to be emptied the level can be progressively lowered by removing the blocks one at a time. Sometimes a third pair of grooves are formed in which a screen can be placed on the pond side of the wooden blocks. This screen prevents the exit of fish.

Figure 9.9 shows a cross section of a typical pond and shows the relationship between the various structures.

9.2   The management of earth ponds

Fish production

The management of ponds is concerned with the water supply and the art of maintaining the environmental conditions required for the optimum growth and minimum mortality of the pond's fish population.

The water enters through a controlled water inlet, the purpose of which is to assure a regular and adjustable water supply, to prevent the escape of fish and to prevent the entry into the pond of unwanted species. The water leaves the pond via a controllable outlet called a monk. The purpose of the monk is to control the level of water, prevent the escape of fish and enable the farmer to draw off bottom water when required to clear any layer of poor quality water which may have formed. The design is dealt with in the chapter on pond construction.

The water supply may come from three different types of supply, spring and well water, rain water and runoff water, and water from a watercourse. Most spring and well water will be of a good quality. However, if water from a natural watercourse is being used care should be taken that there are no sources of pollution immediately upstream. Additionally, it behoves the farmer to pay special attention to the quality of water which is being returned to a natural waterway. If necessary it may be advisable to allow the water to pass through a settling tank or pond which removes pollutants such as faecal matter.

The amount of water required by a pond depends on a variety of factors, the most important being the following: the type of fish being farmed, the temperature of the water, the density of fish kept, the degree of water leakage, and the extent of evaporation from the water surface. In general fresh water is

...

...

...

...

...

...

...

...

...

...

...

...

...

...

...

required for two main purposes, to maintain the oxygen content of the pond water at the required level, and to maintain the required depth of the pond.

The water supply should, whenever possible, enter the pond under gravity and not by pumping, which is expensive and subject to breakdown.

The amount of water required varies considerably as can be seen from the following examples:

Pond cultivation of trout: 5 litres/sec/ha.
Intensive cultivation of trout: 100 litres/sec/ha.
Tropical earth ponds: 3 litres/sec/ha to 12 litres/sec/ha depending on the local conditions of evaporation.

Pond management covers all the operations which are required to keep the pond at maximum efficiency; the engineering required to keep the pond banks and installations in good condition, the maintenance of hygenic conditions, the control of water quality, the control of parasites and diseases, the maintenance of productivity and the control of unwanted growth of vegetation.

Like a farm field, the pond bottom has to be looked after. After each harvest the pond is drained and the bottom is allowed to dry out. In some ponds the remaining fish are kept in a deeper trench at one end which is treated separately. The objective of the treatment of the bottom is to create a layer of fertile soil, in much the same way as a farmer creates a fertile field. When the bottom is dry, excess mud and detritus are removed, the soil is ploughed and then treated with lime and manure as required. The lime corrects the acidity which builds up in the mud and also it helps to eradicate disease and parasites. At this stage other chemicals can be applied to eradicate unwanted species if necessary. The bottom is also inspected for plant roots, which have to be removed before they can damage the structure of the pond.

The water quality is managed by the farmer to maintain the optimum living conditions for the fish. Pond management activities can be divided into two main types: control of the water entering and leaving the pond, and control of the amount of fertilizer added.

Control of the water flow is used mainly to adjust the temperature and oxygen content of the water. This the farmer can do by increasing or decreasing the water flow and, in extreme cases, by drawing down the water level in the pond and replacing the water. When the economics of the farm allow, oxygen content can also be adjusted when too low by the use of mechanical aerators which agitate the water and so increase the oxygen uptake. The acidity and alkalinity of the water can also be adjusted by increasing the water flow but this can be controlled more effectively by the addition of lime.

Control of water temperature is important for two reasons; first, because each species has a thermal range within which it will live and above and below which the animals will die; and secondly, it greatly affects the amount of oxygen that the water contains. The level of dissolved oxygen varies from 14.6 mg/ litre at 0 °C to 7.6 mg/litre at 30 °C. Thus over the temperature range of 0–30 °C the amount of oxygen contained in the water is halved; this has great significance to the fish farmer. The oxygen needs of fish vary with the species. Salmon require 9 mg/ litre; carp 6 mg/litre, but can withstand levels as low as 3 mg/ litre. Tilapia also can withstand low levels of oxygen below 6 mg/litre.

Low oxygen levels in the pond water can have many causes; high water temperature, high population densities, excessive use of manure, decomposing organic matter in the pond and heavy growth of plankton and higher plants. Falls of oxygen due to too high levels of plankton or plants are particularly dangerous for a farmer as these occur at night. This is due to the fact that at night the plants are respiring but not photo-synthesizing and thus their oxygen balance is negative, whereas during the hours of light their oxygen balance is positive.

The acidity and alkalinity of pond water is measured by special apparatus. The acidity of a liquid is expressed as the pH of the liquid, pH being a measure of the concentration of hydrogen ions in the liquid. The detailed explanation of this term need not concern a fish farmer, but he should understand that pH 7 is a neutral liquid, that lower pH is acid and that higher pH is alkaline. The best conditions for a fish pond are a stable pH with a level between neutral and alkaline.

The alkalinity of water is a measure of the calcium and magnesium content. Alkalinity is important as it controls and stabilizes the pH of water. The farmer adjusts the alkalinity by the use of lime. Lime is added when the pH of the water is low, when the alkalinity is low, when the pond bottom is muddy and neglected, when the organic content of the pond is too high owing to over-manuring or neglect and when there is a threat of disease in the pond. Lime can be applied directly to the water, on the pond bottom when the pond is dry or into the water inflow. If the water is limed then 200 kg/ha of calcium oxide is applied, 200–400 kg/ha being used on the pond bottom. This application is increased to 1000 kg/ha when treating a pond bottom for the eradication of fish parasites, and 1000 kg/ha can be applied to the pond bottom when combating conditions of low pH in the pond.

The productivity of a pond is a measure of the level of production of the organisms which are the feed for the fish. This production level depends on the water temperature, the oxygen content and a required level of essential nutrients which are supplied by the farmer in fertilizers and manures.

Mineral fertilizers are generally applied when the pond is dry or just after filling with water. Two main types are used, phosphate fertilizers and nitrogenous fertilizers. Phosphate fertilizers are applied at the rate of 30 kg/ha of P_2O_5. Nitrogenous fertilizers can be used with phosphate fertilizers, the best ratio of P:N being 1:4.

These fertilizers are used at the rate of 60 kg superphosphate to 60 kg ammonium sulphate/ha.

Manure is applied at the rate of 20 000 kg/ha. It is placed on the bottom of the pond in small heaps. This method of application is important because if the manure is spread over the whole of the bottom it will form a layer without oxygen and will then cause problems. Chicken, cow and pig manure can be used. The fertility of the pond is maintained by further applications of manure. The amount used has to be worked out by the farmer; too little causes reduced fertility and too much causes fouling of the water.

Organic fertilization can be obtained directly by farming animals in conjunction with fish ponds. Pigs may be farmed next to the pond and the manure washed directly into it or ducks may be kept on the water and so manure the pond continuously. Ducks are generally farmed at a density of 250 ducks for every hectare of pond, and pigs are generally kept at a ratio of 200 per hectare of a fish pond.

Shrimp production

As in fish farming the objective in shrimp farm pond management is to ensure that the animals in the pond are kept at optimum environmental conditions.

Water replenishment is achieved in one of two ways; sea water is either pumped into the pond or allowed to enter the pond during the tidal cycle. Pumping the water gives the farmer a high level of control over the water supply and, with filtration, it allows the farmer to exclude all unwanted animals. Pumping is expensive, involves a high capital outlay and means that the water supply is completely dependent on the pumps. Water exchange with tidal action is inexpensive but is much more difficult to control and makes the exclusion of unwanted organisms very difficult. Tidal water exchange can be used only where tidal level differences are adequate over the whole year to accomplish water exchange. In regions subject to severe storms a further hazard is the flooding of ponds situated below high tide level as is necessary for tidal flushing. When tidal flushing is used, up to 80% of the water can be exchanged during the tidal cycle. When the water is pumped the farmer can decide the exchange rate.

Shrimp grow best within a temperature range of 25–30 °C. They require water fully saturated with oxygen and low in hydrogen sulphide. They are adversely affected by concentrations of 0.1–2.0 ppm of hydrogen sulphide in the water and 4.0 ppm is lethal to them.

The pond bottoms are dried out after harvesting and are then treated with derris root or tea seed cake to eradicate any fish present. The pond is also cleared of unwanted seaweed, organic deposits and mud. The pond bottom is often cleared with bulldozers which are then used to turn over the top layer of soil. After clearing, the pond is covered with a shallow layer of sea water which is then left to allow the layer of lab-lab to become re-established.

After the pond is filled and re-stocked, management is concerned with three main objectives, to maintain the required level of feed for the animals, to maintain the level of oxygen in the water and to maintain a low level of hydrogen sulphide. Oxygen levels are maintained by flushing with new water and, if necessary, by the use of mechanical aerators. Concentrations of hydrogen sulphide are reduced by treating the pond with iron oxide which is broadcast over the pond at a rate of $1\,kg/m^2$. When water temperatures are high this application is often repeated twice daily.

The management of ponds for freshwater prawns is essentially the same as for sea water shrimp except, of course, that these hold fresh water and the prawns will tolerate only salinities of up to 10‰. High oxygen levels have to be maintained and mechanical aerators are often used for this purpose. The water in the ponds is exchanged at a rate of 140 litres/min/ha of pond.

Chapter 10
The Supply of Aquatic Seed

In aquaculture the word 'seed' denotes the young stages of the animal used for stocking farm ponds and with which a farmer starts the production cycle in his farm.

10.1 Fish seed

The provision of seed for fish farming falls into two categories resulting from an important difference between the spawning behaviour of various species. This is that whereas some species will spawn in farm ponds naturally and readily, and so can be easily made to provide fresh seed, others will not spawn naturally under farm conditions. In the latter case the seed is either collected from the wild or produced by inducing the fish to spawn under farm conditions.

Probably the best known fish that spawn easily in a pond are tilapias. They can be produced in two ways; either they are allowed to spawn directly in the production pond in which they are living, or, in more controlled systems, the fish spawn in one pond and the fry are then removed at intervals and transferred to a nursery pond before being moved to the final growing pond.

A variation of the production system for fish which mature in the pond is used for trout. At maturity the trout, in some circumstances, will spawn in the pond. However, owing to the requirements of the trout eggs and the use of concrete ponds, the eggs will not hatch or, if they do, the fry will never be reared under pond conditions. Mature trout are, therefore, removed from the pond and spawned by hand by exerting pressure on the female trout's body to remove the eggs, and by mixing these eggs with milt (sperm) which has similarly been extracted from a male fish. The fertilized eggs are then hatched and the young reared in a hatchery.

Carp form an example half-way between the tilapia and the trout. Carp will spawn in a farm pond and the eggs and fry can be reared in the pond. The eggs are laid on objects such as grass, leaves or tree branches in the water; after spawning the adult fish are removed and the fry reared in the pond. Alternatively the eggs may be removed along with the object to which they are attached and transferred to a hatchery or another pond for rearing.

Many of the commercially important species used in fish farming will never spawn in ponds. This single factor has been one of the principal restraints to the development of fish farming in many parts of the world. For example, Chinese and Indian carp, among the main species in aquaculture, will not spawn in ponds; the commercially valuable river fish of South America have to complete migrations of hundreds of kilometres before they can spawn; mullet and milkfish both have to carry out migrations at sea before they are ready to spawn. In order to obtain seed for farming such types, farmers have two choices; either they can gather seed from the wild or they can induce the fish to spawn under farm conditions.

The gathering of the seed of these species from the wild is practised commercially in many countries. Nevertheless the use of a supply of seed from the wild as the basis of an industry has many dangers as the supply is dependent on many factors out of the control of the farmer. For example, wild populations of animals are subject to great natural fluctuations in their numbers from year to year; pollution, especially oil pollution, can greatly decrease the number of fry in any one year; the system is very wasteful as the fry are very delicate and many die during transfer to the farm. Apart from these dangers the use of wild seed means that it is impossible for the farmer to improve his stock by selective breeding as land farmers can. For these reasons great efforts have been made to develop technologies which will enable the seed of species which will not spawn in ponds to be produced on the farm.

The basis of such systems is the use of hormones from the pituitary gland; these hormones, as was explained earlier in the description of endocrine organs, control the maturation of the fishes' gonads. Treatment with these hormones causes the fish to become ripe without having to undergo spawning migrations. Adult fish nearing sexual maturity are given injections of either whole fish pituitary glands or of pituitary hormones and the injections cause the fish to be ready to spawn. The fish are then spawned by manual pressure and the resulting eggs hatched in a hatchery. This technology has now been perfected for many species.

Mention should be made of the development of the technique of cryopreservation in which viable fish sperm can be stored at very low temperatures for long periods. The use of this technique enables sperm to be stored and transported as required, with obvious benefits for breeding programmes.

Unfortunately this is not the end of the problem of producing seed of these species, most of which produce small eggs which hatch into very small fry with a very small food reserve. Great problems have been encountered in rearing these small fish at an acceptable mortality and growth rate. In nature the little fish feed on different micro-organisms present in the water in which they are living. The difficulty has been to produce these organisms in culture and to supply the correct environmental conditions for the fish. For many species, however, these problems have now been solved.

The farm hatching of eggs and the rearing of the very young stages of the fish have necessitated the development of hatcheries for the eggs and rearing facilities for the young stages. The facilities range in complication from the simple systems used for salmon and trout eggs, which hatch with large yolk reserves, to complicated systems used for mullet culture where different species of micro-organisms have to be cultured to provide food for the developing stages of the fish. These hatchery and larval rearing facilities are dealt with more fully in Part 3 of this manual.

10.2   Shrimp seed

In the past shrimp farms everywhere depended on the wild production of young shrimp for their seed supply and in some countries this is still the case. The collection of these young stages is made possible by the fact that they migrate into coastal waters where they seek the rich feeding in areas such as mangrove swamps. In some countries, however, the growth of the shrimp farming industry with an ever-increasing demand for these seeds, together with a reduction of the nursery bed areas due to reclamation of the land and pollution, has resulted in a shortage of shrimp seed.

To combat the shortages and the unreliable annual fluctuations in the quantity of natural seed, hatchery systems for shrimp have been developed. At present all shrimp hatcheries, with the notable exception of one or two highly sophisticated units, use mature wild females for the supply of eggs. As the mature female carries its own sperm package only females have to be captured for the hatchery. The females are caught at sea and transferred to tanks where the eggs are shed by the females and fertilized. The eggs hatch in the hatchery tanks and are then reared to stocking size before transfer to the farm ponds. This process has been taken a step further in a very few farms where the technology for the maturation of the males and females in the farm ponds has been developed. These farms are therefore no longer dependent on the supply of mature females from the wild, which in many countries was becoming difficult and the females very expensive.

In the case of the freshwater prawn the production of ripe females on the farm and the rearing of the young is a well-established technology and many commercial farms produce their own seed. In some countries, however, the seed is still collected from the wild by the farmers. The hatcheries for freshwater prawns have to be supplied with sea water as the developing young require water of a salinity of 12 to 16 parts per thousand.

10.3   Bivalve mollusc seed

Most of the seed for the world's oyster and mussel farms is collected from the wild by the farmers. However in Europe, Japan, the United States of America and elsewhere a shortage of oyster and other bivalve seed, called spat, together with the very high market price and the good market demand, especially in Europe, has led to the development and establishment of commercial hatcheries which are now trading on a world-wide basis.

Chapter 11
The Feeding of Aquatic Farmed Species

11.1   Feeding fish

The type of feeding regime used in a fish farm can be classified
into three different systems: either food is produced directly in
the pond in which the fish are living by the plants growing in the
pond as a result of fertilizing; or food is produced in the pond,
but also, because of the high density of the fish population,
supplementary food, produced outside the pond, is given; or all
the food given to the fish is produced outside the pond.

Food produced totally in the pond

In this system the sun's energy is converted by the plants in the
pond using required nutrients from the manure added to the
pond.
  In terrestrial agriculture a herbivore, such as a cow, eats the
grass which grows in the field; in aquaculture a fish eats the
'grass' which is produced in the pond. In a pond this 'grass' is
composed of millions of tiny animals and plants and is called
plankton. The basis of the food production is therefore the same
as the basis of all food production, that is energy from the sun.
In the pond this energy is used by the huge population of
tiny plants living in the water to produce carbohydrates. This
floating population of plants and animals occurs both in freshwater ponds and in the sea. The plants are in fact the smallest
true plants that we know; these are called algae and they are
responsible for the green colour which can be seen in the water
in fish ponds. The tiny animals living on the plankton are often
the larvae of insects as well as tiny animals which complete their

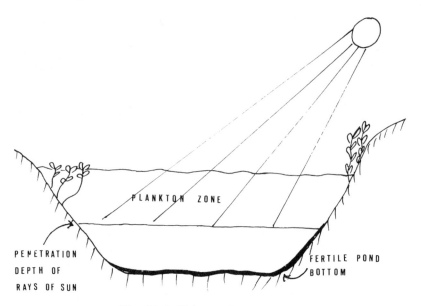

PLANKTON ZONE

PENETRATION
DEPTH OF
RAYS OF SUN

FERTILE POND
BOTTOM

Fig. 11.1 Fish pond zones.

life cycles in the plankton. In sea water plankton also contains the larvae of crabs, lobsters, shellfish and fish.

In shallow ponds the plankton may extend to the bottom of the pond where other types of animals and plants form a carpet over the bottom. The depth to which the plants grow in the water body depends on the penetration of sunlight, as the plants require the sunlight in order to perform photosynthesis, which is the production of carbohydrates using the energy of sunlight. The penetration of the light depends on the clarity of the water, so water in which there is a dense growth of plankton will have light penetration to a shallower depth than a pond in which there is little growth. Dense plankton in deeper ponds can prevent the growth of plants on the pond bottom by preventing sunlight penetration. This factor is extremely important. A successfully managed pond has the optimum level of plankton at all times combined with the growth of plants on the pond bottom.

In order to grow, plankton, like any other plant, requires fertilizers. As on land these fertilizers can be supplied either as animal manure or as artificial fertilizers. Artificial fertilizers are expensive compared with manure so that they are less likely to be used when the market price of the fish produced is low or where they are produced for home consumption. As on land, manure has to be used with care in a fish pond. The application of too little manure does not achieve the desired effect, and the use of too much fouls the water, uses up oxygen, stops growth of the desired species of plankton and can even kill the fish.

The management of the pond is the process whereby the desired types of planktonic organisms are maintained at the optimum density required throughout the growing season of the fish. To control plankton production the farmer takes samples of water at intervals to determine the plankton concentration and ideally to obtain some idea of the plankton composition. These measurements, together with measurements of the oxygen content of the water, its temperature and pH, allow control to be maintained. For instance the farmer may find that the water is becoming too acid; then it is necessary to add lime to the water to make the water alkaline or nearly neutral. Phytoplankton requires a pH of about 7.0 and a slightly alkaline pH is best for zooplankton.

The young fish and some specialized adult fish can feed directly on the plankton. For example one litre of rich plankton will feed about 4000 young fish 1.5 to 2.0 cm in length for about eight hours.

To ensure the optimum production of plankton regular fertilizing is needed: thus the sun's energy is converted into carbohydrates by the planktonic algae, the planktonic animals feed on the bacteria in the plankton and to some extent on the plants.

Some fish feed directly on the plankton, uneaten plankton dies and fertilizes the pond by forming debris on the bottom which is eaten by other fish species. The uneaten debris fertilizes larger plants which in turn are eaten by other fish species. Larger invertebrate animals such as snails and shrimps feed on the debris and on the larger plants and are eaten by other fish species.

The variable feeding habits of different species are of great importance to the fish farmer. By growing fish with different feeding habits in the same pond a more complete use of the food production of the pond is possible. In the final analysis the production of a fish pond is a measure of the efficiency of the use of the energy of the sun that falls on the pond. A good fish farmer can achieve a higher output by careful pond management.

Use of supplementary feed

In this system the fish obtain their food partly from the natural production in the pond and partly as supplementary feed which is thrown into the pond by the farmer. This supplementary feed is required when the fish are stocked in the pond at such a high density that sufficient food cannot be produced naturally to meet their requirements. The supplementary feed acts in two ways; it is used directly as food but also acts as an additional fertilizer. It has recently been shown that this second action is more important than was earlier believed; research work has demonstrated that, even in ponds receiving a high level of supplementary food, some 50% of the body weight of the fish is derived from the natural food chain in the pond and not directly from the supplementary food.

Generally local waste products are used as supplementary feeds, the type used depending on the local availability and costs. Typical examples of supplementary feeds are rice bran, cereals, fruit, vegetables, brewers' wastes and cereal wastes. In addition, manufactured fish feeds can also be used but as these are expensive, their use can be uneconomic, depending on the farming system and the species being farmed.

Total use of manufactured feed

When fish are reared totally on food which is added by the farmer the economics of the system have to be carefully established. This type of farming is therefore found mainly in the production of fish with a high market price which are farmed at high densities such as salmon, trout and catfish; it is also used occasionally for farming tilapia, but in these cases the tilapia are reared at high densities in carefully controlled farm conditions. The great advantage of using manufactured fish feeds lies in the fact that, for fish such as salmon and trout, the nutritional requirements of the fish are known in great detail and therefore the farmer can be sure that his fish are receiving a proper diet. These modern feeds are produced in the form of dry pellets which can be stored and used as required; automatic feeding systems can also be used to deliver the pellets into the ponds, thus reducing labour costs – an important factor.

The high price of the feed is in part due to the high price of its protein content. Manufacturers are continually carrying out research to try to reduce these costs by the use of a variety of protein sources. At present the main source is fish meal, which is now a very expensive product; other meals are being tried such as blood meal, chicken feather meal and a meal made from bacteria grown on cereal wastes or oil by-products.

11.2   Feeding shrimp

Three basic types of feed are used in shrimp farms: natural feeds, formed on the bottom of the pond and in many countries called lab-lab – they are composed of a thick layer of algae, small animals and detritus; prepared natural feeds – such as minced fish offal or shellfish; and formulated manufactured feeds.

Dependence on lab-lab as the sole feed for the shrimp is found in artisanal shrimp farms. Such a feeding system will support only low population densities and hence produce low yields. However lab-lab plays an important role in the other two feeding systems used. Without lab-lab formation productivity would not be as high as with prepared natural feeds. Formulated feeds, although eaten by the shrimp, also serve as fertilizers and encourage the production of lab-lab, which is then eaten by the shrimp. The lab-lab therefore increases the efficiency of utilization of the feed by the shrimp.

11.3   Feeding bivalve molluscs

There are no bivalve farms which use any added feed for adults. Some research work is being carried out to try to develop closed cycle farms using added feeds but at present farms rely on the food brought by the open sea. Larvae and seed produced in hatcheries are fed on specially reared micro-algae.

Chapter 12
Harvesting

12.1 The harvesting of fish

As in any other type of farming the final phase in the fish farming cycle is the harvesting and sale of the produce. Although marketing is beyond the scope of this book, it should be realised that it would be foolhardy to enter into large scale aquaculture production without having first considered very carefully the anticipated market demands and outlets. Furthermore, marketing also requires careful assessment of the optimum product type (fresh, chilled, frozen, smoked, canned), methods of post-harvest storage, availability of transport, packaging requirements and the seasonal variations in supply and demand.

There are two ways in which a farmer can harvest his product; he can either take out the whole population from a pond at the same time or he can selectively cull fish from the pond throughout the year. In the latter system it is usually the larger fish that are taken for market and the smaller fish are left in the pond to grow. It is, of course, possible to combine the two methods by taking large fish as required and finally removing all the remaining fish at one time.

The method used for continuous culling depends on the fact that if a net is hung in a pond the fish will attempt to swim through the meshes of the net. By selecting the aperture size of net mesh the farmer can make sure that any fish smaller than he wishes to harvest will swim through the net whilst the larger fish will get stuck. Fish are often caught by the operculum, and because of this the net is called a gill net. In this way it is possible to harvest fish throughout the year without having to draw down the water in the pond or disturb the unwanted fish in any way.

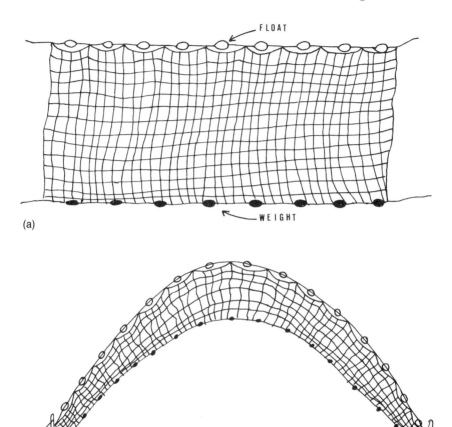

Fig. 12.1 (a) A gill net. (b) A seine net.

When all the fish in the pond are to be harvested at the same time it is necessary to lower the water level to be sure that all are caught. The problem then is to ensure that the fish are harvested in good condition and in a reasonable time. For this reason it is usual to use two catching systems. First the majority can be caught in a seine net whilst the level of the water is high. A seine net is a length of net with a small mesh size; it has floats along the top and weights along the bottom. The net is paid out from the bank of the pond and pulled in a semicircle until it

Fig. 12.2 The use of a seine net to harvest a pond on a catfish farm in the USA.

reaches the bank again; the net is then dragged into the bank, trapping the fish. As the pond is emptying large quantities of fish are caught in this way. At the same time fish leaving the pond in the outflowing water are caught at the exit of the pipe. Slatted boxes or nets are placed under the water outfall to catch the fish. When the pond is empty all the remaining fish have to be gathered up from the mud on the pond bottom by hand. For this reason it is desirable to catch as many of the fish as possible before the pond is emptied of water as fish left in the mud can be missed or damaged.

In a large pond the harvest will result in a heavy weight of fish which have to be lifted out of the pond and into a transporter. Two different types of mechanical lifts have been evolved for moving the fish. In the first system the fish, as they are gathered in the seine net, are pumped out in a mixture of fish and water. Special pumps are available for this work. In the second system the fish are lifted from the seine net by a mechanical hoist, which consists of a hopper and a system of lifting the hopper. The hopper is filled in the pond with a mixture of fish and water, lifted up and emptied into the transporter.

After harvesting, the pond is dried out and then treated to kill

Fig. 12.3 The use of a mechanical hoist to empty the catch from the seine net.

unwanted animals and plants and to eradicate disease organisms.

12.2   The harvesting of shrimp

The major problem encountered when harvesting shrimp is caused by their habit of burrowing in the pond bottom; this means that an ordinary seine net will pass over the majority of the animals.

Two ways have been developed to overcome this problem. In the first system the migratory habit of the shrimp in the pond is utilized. The moving shrimp are directed by a series of nets set at angles to the edge of the pond, so that they are directed into conical nets where they are trapped for harvesting. In the second system the shrimp are disturbed from their burrows by a series of water jets which move across the pond in front of a seine net. The shrimp rise up in the water and are trapped in the net. In this system some selection for size is possible as the smaller animals can pass unharmed through the net meshes.

Freshwater prawns do not burrow and therefore can be harvested with a conventional seine net. Again it is possible to harvest only the larger animals by use of nets with a selected mesh size.

12.3   The harvesting of molluscs

Harvesting of immotile bivalves is relatively easy. Suspended ropes and cages are raised, either manually or mechanically. Littoral cages and trestles are serviced at low tide.

Part 3
Farming Methods for the Principal Species

This part of the manual deals with actual farming operations, the object being to give a concise account of the operational procedures of farming systems for the principal species currently farmed. The synopses are outlines of the farm processes so that the reader may have a clear idea of what is involved. It is not supposed that these synopses can form a blue-print for a farm. As in all agricultural operations a basic design has to be adapted to local conditions. Furthermore it is difficult to obtain precise information about many aquaculture operations which are understandably the subject of commercial secrecy.

No detailed descriptions are given of the various constructions and apparatus used but more information can be found in the literature cited in the bibliography.

The species described are those currently farmed in established farming procedures. These often differ according to country but the basic farming operations are the same.

Chapter 13
Salmonids

The family *Salmonidae* contains freshwater and anadromous species. Anadromous species migrate from the sea to spawn in fresh water, and catadromous species migrate down rivers to spawn in sea water. *Salmonidae* occur naturally in the northern hemisphere but they have now been introduced into the southern hemisphere. There are about 68 species of *Salmonidae* which have been divided into three subfamilies. Of these, the subfamily *Salmoninae* contains all the species of interest to fish farmers. These are classified as follows:

Genus: *Salvelinus*;
   e.g. Salvelinus alpinus – Char.
   e.g. Salvelinus fontinalis – Brook trout.
Genus: *Salmo*;
   e.g. Salmo trutta – Brown trout.
   e.g. Salmo gairdnerii – Rainbow trout.
Genus: *Oncorhynchus*
   e.g. Oncorhynchus nerka – Sockeye salmon.

Of these the one of most importance in fish farming is the rainbow trout. All species are farmed, but only rainbow trout, Atlantic salmon (*Salmo salar*) and Pacific salmon species are farmed for food.

Salmonids all require water of high purity with a high oxygen content and temperatures below about 20 °C. The optimum temperature for growth varies between 12 and 20 °C, depending on species. They show a wide variation in their resistance to changes in water salinity. This resistance varies between species and between development stages of the same species. The three most important salmonids from the farmers viewpoint, the rainbow trout, the Atlantic salmon and the Pacific salmon, all change their ability to withstand high

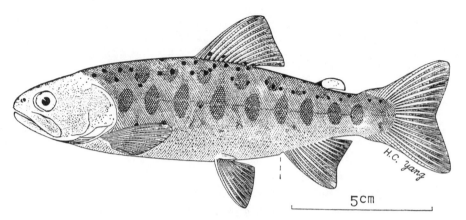

Fig. 13.1 Rainbow trout (*Salmo gairdnerii*) (from Chen: *Aquaculture Practices in Taiwan*).

salinity water as they grow. All spawn in fresh water and the young stages develop in fresh water. Rainbow trout and Atlantic salmon become able to withstand seawater salinities after perhaps 6 months of life in fresh water; Pacific salmon can enter the sea at very early development stages. These differences play an important role in the commercial farming of these species.

All salmon species possess a remarkable homing instinct which allows them to return to the river in which they grew up when they become sexually mature. This fact is made use of in the development of the ranching system for their production.

All salmonids are carnivorous and so require a high percentage of protein in their diet. This makes the price of feed high and would make salmonid farming economically difficult if it were not for the high market value of the product. The high cost of feed adds another advantage to the ranching system.

13.1   Egg production

Egg production presents few problems to the farmer. Trout will readily become ripe in the ponds and salmon can be captured when they are ripe as they ascend the river to spawn.

The fish lay large eggs about 0.5 cm in diameter. The number of eggs produced depends on the size of the female, the average being about 1500 eggs for every kg of fish body weight.

The females are stripped of their eggs by hand. The hand of the operator is run down the fishes' abdomen expelling the eggs into a receptacle where they are fertilized by milt which is added to them, this having been obtained by manual pressure on the male fish.

13.2 Hatcheries

The fertilized eggs are developed in hatcheries on the site. These are of two different basic designs. The eggs are hatched either in troughs of flowing water or in incubators down which water trickles. In the troughs they are housed in trays with baffles along the front to ensure that a water circulation is maintained between the eggs. A tray 50 cm × 50 cm will house between 10 000 and 15 000 eggs, depending on their size. In incubators the eggs are housed on trays which slot into commercially-produced incubators. Each tray is about 35 cm square in area and will house between 30 000 and 70 000 eggs, depending on their size. Incubators will hold a number of trays and one holding 20 such trays will require a water-flow of 14 litres/ minute. The water trickles down from the top of the incubator over the eggs in the stacked trays. In all cases incubating eggs have to be kept in the dark as daylight will damage or kill the eggs.

The hatching time of the eggs varies with the water temperature and with the species. It is calculated by farmers as degree days, the number of days for hatching multiplied by the average water temperature (°C) over the period. Rainbow trout take 290–330 degree days and brown trout take 400–460 degree days.

The young fish hatch from the egg with a large yolk sac and do not require to feed until this yolk has been used up. These young are called alevins and this stage lasts about 220 degree days for brown trout and 180 for rainbows. During the alevin stage the fish are kept in the dark.

After absorption of the yolk sac the young fish are called fry. The young of the Pacific salmon species are capable of entering sea water at the fry stage or even a little earlier.

13.3   Fry production

The fry of the fish are kept in daylight and are reared in troughs, either those which housed the egg trays or special fry troughs.

The fry are raised at densities of between 10 000 and 30 000 per m^2 and with a water flow of 1–2 litres/min/1000 fry.

13.4   Growout facilities

As the salmonids require a diet high in protein earth ponds are less frequently used for their production. At the water temperatures required for the growth of these species, protein production in the food chain of an earth pond would be very low and therefore would support only a low production. The fish are therefore usually reared in fabricated ponds or floating sea cages and fed prepared food. Basically the main types of ponds used are concrete raceways, concrete ponds or circular fibreglass ponds.

Raceways are narrow concrete ponds in which a frequent interchange of water is possible by maintaining a fast rate of flow down the length of the narrow pond. A typical dimension for such a pond would be 30 m long, 3 m wide and 1 m deep. Concrete ponds are generally rectangular, a typical example being 25 m × 10 m × 2 m deep.

Then there are circular ponds. The advantage of this shape of pond is that if water is jetted into the pond at an angle from the side and allowed to exit from the centre via a stand pipe, the water body of the pond will rotate round the pond. This swirling of the water allows a high density of fish to be kept in these ponds and also all debris collects in the centre from where it can more efficiently be removed.

The numbers of fish that can be grown in any of these facilities depend on three factors, the size of the fish, the water temperature (this is presuming that the water is fully saturated with oxygen) and the water flow. Five litres of water per min will support 1.25 kg of 5 g fish at 20 °C or 10 kg at 7 °C, or 2.5 kg of 500 g fish at 20°C. A general calculation for the potential production of a farm can be based on the ratio of 1.5 kg/year/litre/min.

The carrying capacity of a circular pond is high. For example, a 4 m diameter pond with a water flow of 250 litre/min will carry a population of 200 kg of trout.

A growout facility for salmon which is now extensively used is the floating sea cage. Such cages are particularly well suited for production in sheltered waters such as those found along the fjord coastlines of Norway, Scotland, British Columbia and southern Chile. These areas frequently have a tradition of sea fishing for salmon and cage production is now forming the basis for the continuity of fish supplies in the wake of stock depletions and increased market demand. Big sea cages for salmon have a capacity of 700 m^3 and are stocked at densities of 30–40 kg per m^3. Fish are either hand- or automatically fed with dry pelleted feed or wet trash fish. Considerable engineering capital and maintenance costs are required for this type of facility and extra costs have frequently been incurred following cage damage and the resulting loss of fish. Fouling of the netting is a major problem and many farms use trained staff in scuba gear to keep the netting walls clean and in good repair.

13.5   Ranching

In ranching salmonids, all of which have the remarkable facility of returning to the river of their birth when they become sexually mature, both the pumping costs and the costs of feed are saved. Capital costs, feed costs and maintenance costs are thus all kept to an absolute minimum. The system has, however, a disadvantage in that the farmer has no control over his adult stock, which have no protection from predators. Returns of up to 20% are reported and the ranching of salmonids has now developed into a major industry. Ranching, or the related practice of stock enhancement, is carried out in Alaska, Washington State, Oregon, the Pacific Coast of the former USSR, Japan, the Atlantic Provinces of Canada, Iceland, the Baltic, Chile, New Zealand, Norway and China.

13.6   Feed

Most salmonid farms now use manufactured dry feed. These pelleted feeds are produced to a formula which provides all known dietary requirements of the fish, both quantitative and qualitative.

The use of this diet allows the farmer to achieve very high conversion ratios, the conversion ratio being the measure of the weight of feed required to produce a unit weight of fish. In commercial operations conversion ratios of 1 : 1.5 are common. It must be remembered that this is dry weight of food to wet weight of fish.

The modern intensive trout farm is directly comparable in efficiency and commercialization to the terrestrial modern poultry farm. Indeed conversion ratios for fish are often better than for chickens, largely because the fish, being cold-blooded, does not have to use energy in maintaining a body temperature higher than that of its environment.

Chapter 14
Catfish

Catfish belong to the order *Siluriformes* and are among marine or fresh-water species found in most parts of the world. Over 2000 species have been recorded, of which over half are in South America, but at present only one family is used for large-scale commercial farming, the North American catfish family, the Ictaluridae. Attempts are being made to farm some of the South American catfish but as yet there is no significant farming of them. There is, however, a long tradition of catfish farming in the Far East, where members of the *Pangasiidae* and the *Clariidae* are farmed.

Family: Ictaluridae; Species *Ictalurus punctatus* and *Ictalurus furcatus*, the channel catfish and the blue catfish farmed in the USA.

Family: Pangasiidae; Species *Pangasius sutchi* farmed in Thailand, Cambodia and Vietnam; *Pangasius larnaudi* and *Pangasius sutchi* farmed in Laos and India.

Family: Clariidae; Species *Clarias batrachus* and *Clarias microcephalus* farmed in Thailand.

All the catfish used in farming are freshwater species. Catfish do not have their skin covered by the usual scales but have either a naked skin or one covered with bony plates. This is a useful factor for the farmer as it means that catfish can be handled easily without the danger of rubbing off scales and so damaging the skin.

Catfish spawn readily in ponds in shallow water where the eggs are laid in a nest and guarded by the male.

Catfish are omnivorous bottom feeders. They are warm water fish with a temperature range of 16–30 °C.

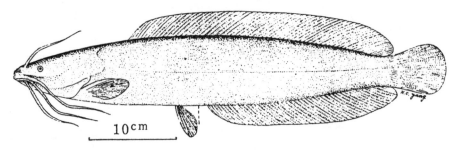

Fig. 14.1 Walking catfish (*Clarias fuscus*) (from Chen).

14.1 Egg production

Egg production in catfish presents few problems as the fish will spawn in farm ponds. In North American catfish farms spawning is arranged using one of three different systems, a pond in which has been placed a container suitable for a nest site, a pen constructed in the pond, or a hatchery tank.

In pond spawning, paired catfish are left in the pond which contains a suitable nest area in which they can spawn. Spawning ponds are about 0.25 ha in area and are stocked at a density of between 50 and 300 fish per ha. In pen spawning each pair of fish is confined with a suitable spawning container in a wire mesh pen 3–6 m^2 and 1 m deep. In both systems the eggs may be left to hatch in the pond or may be removed for hatching in a hatchery. Spawning in a hatchery tank is induced by injecting the females with three doses of acetone-dried fish pituitaries, with an average weight of 10 mg, or with one dose of human chorionic gonadotropin, containing 60–2200 International Units; the males are not injected. The resultant eggs are hatched in the hatchery. Females lay between 3000 and 20 000 eggs, depending on their size.

In the case of the Pangasiidae and Clariidae a high proportion of the seed is obtained from the wild as fry. However, some induced spawning is now carried out in Thailand. Clariidae will also spawn in farm ponds.

14.2   Hatcheries

The eggs are hatched in simple hatcheries consisting of troughs of running water in which the egg masses are placed, the egg masses being kept in motion in the water by means of mechanical agitators.

Ictaluridae eggs hatch in 5–10 days at 20 °C and the eggs of Pangasiidae in 1–3 days at 28 °C.

14.3   Fry production

Catfish eggs are small and hatch into very small larvae. Channel catfish larvae hatch with a very small yolk sac. The fry are reared in nursery troughs until the yolk sac is absorbed and the fry have started to feed. Nursery troughs are 3 m long, 50 cm wide and 30 cm deep and they are supplied with water at a rate of 20 litres per minute. The fry are stocked at 10 000 per trough.

The fry start to feed about 4 days after hatching. Fry ponds vary in size about 0.5 ha and are stocked with feeding fry at a density of 100 000 per ha. The fry are fed pelleted food and their mortality during this stage is about 35%.

Pangasiidae fry are generally stocked directly into the fry ponds after hatching, although some use is made of fry troughs. The fry feed on the results of the natural productivity of the pond.

14.4   Growout ponds

These ponds vary in size between 0.5 and 2 ha. Because of low winter temperatures which slow down growth, channel catfish are sometimes grown over 2 years to produce market size fish. During the first year the stocking density is about 20 000 per ha, which is reduced to 4000 during the second year.

Growout ponds for Clariidae and Pangasiidae vary in size between 0.1 and 2 ha and have a depth of 1–3 m. Fry are stocked at a rate of 250 000 per ha.

In Thailand and in Cambodia catfish are also produced in floating netting cages which vary in size between 6 and 100 m^2.

14.5   Feed

In North America all catfish farms use pelleted manufactured rations formulated to meet the known dietary requirements of the fish.

In the Far East the catfish are fed by the productivity of the earth ponds in which they are kept, which is augmented by fertilization. The fish in floating cages are fed with trash fish and plant wastes.

In North America production levels of 2000 kg/ha per year are obtained. In the Far East local catfish have a higher growth rate; in cages the fish reach 1 kg in 10 months and in the ponds 150 g in a year. Production levels are high with production rates up to 97 000 kg/ha per year, produced in three crops, being claimed.

Chapter 15
Tilapia

Tilapia are predominantly warm- and freshwater fish native to Africa. They are a robust fish able to withstand high water temperatures and low levels of oxygen. Their optimum temperature range is between 20 and 30 °C; they can tolerate temperatures as low as 12 °C. Many species can withstand saline water up to full sea water.

Tilapia species have biological differences associated with their feeding and breeding habits and anatomical differences associated with their feeding habits. These differences have caused systematists to divide tilapia into three different genera:

Tilapia, example *Tilapia zillii*. These spawn in nests which they construct in the pond bottom. They guard the young in the nest. They have coarse teeth and feed on macrophytes.
Sarotherodon, example *Sarotherodon galilaeus*. These take the eggs and larvae and brood them in their mouths. they have fine teeth and feed on particulate matter.
Oreochromis, example *Oreochromis spilurus*. These spawn on the ground but brood the eggs and larvae in the mother's mouth. They have fine teeth and feed on filamentous and unicellular algae.

Tilapia are fast growers under good conditions, reaching 150 g in weight in 10 months. In the wild tilapia can grow to be large fish, reaching sizes of 1300 g.

Tilapia become sexually mature at an early stage in their development when they reach a size of about 15 cm generally at about 6 months old. After reaching maturity they spawn every 6 weeks. This high fecundity of tilapia has caused the major problem in tilapia farming, which is the problem of overpopulation of the pond. This means that food becomes in short supply and the fish cannot grow. This in turn means that the

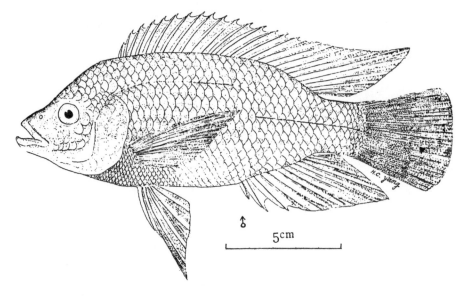

Fig. 15.1 Java tilapia (*Tilapia mossambica*) (from Chen).

farmer when he harvests the pond collects a large population of small fish often too small for the market, and, except in certain places where there is a tradition of eating very small fish, unsuitable for home consumption. Improvements in farming techniques of tilapia have therefore been concentrated in developing ways of overcoming the problem of over-population. Various methods have been developed.

The simplest method is continuously to harvest the fish using a selective seine net so that the largest fish are always removed, thus keeping down the size of the population by removing the market-sized fish. This method is not very effective and is of only limited value in artisanal farms.

A slightly more sophisticated system is to remove the young from the pond as they are produced, rear them in fry ponds and then stock them into growout ponds. Again the fish tend to breed before they are of market size and over-population can result.

The over-population can be controlled by the use of a predator fish. In this system the tilapia are stocked from a fry pond into the growout pond, and when they have reached a size which is too large for them to be eaten by the predator a predator

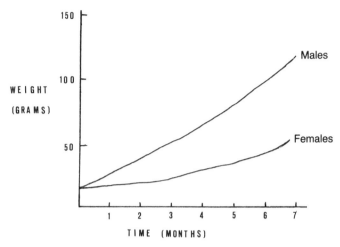

Fig. 15.2 Growth of male and female tilapia (from Chen).

fish is introduced into the pond. This fish will then eat the majority of the tilapia fry when the adults start to breed and so will prevent over-population. Various predators are used in different parts of the world; *Cichlasoma manguense*, El Salvador; *Hemichromis fasciatus*, Zaire; *Lates niloticus*, Nigeria; *Micropterus salmonides*, Madagascar; *Bagrus docmac*, Uganda. The correct balance between the predator and the tilapia is not easy to maintain but this method is an improvement over the straight production systems.

The most effective method of population control is to grow a unisex population so that there can be no breeding. All-male populations are used because males grow faster than do females.

There are three methods by which unisex populations can be achieved. In the first, the fish are sexed by hand when young and the females are discarded, the difference between the genitalia of the sexes being the criteria used for sorting. This system is slow and requires skilled operatives; otherwise a few female fish will be added to the population and breeding will occur.

In the second method all-male populations are achieved by the use of hybrids. It has been found that if certain species of tilapia are cross-bred the resulting offspring are all males.

For example male *Sarotherodon nilotica* crossed with female *Sarotherodon mossambica* give all-male populations.

In the third system the young fish resulting from normal breeding are treated at an early age with male sex hormone. This causes a sex reversal in the young females, resulting in an all-male population. Both methyltestosterone and ethyltestosterone result in sex reversal of the females. The hormone is fed at the rate of 30 µg/g of feed which is fed at 4% of the body weight of the fish for three weeks then at 3% for 17 weeks. The treatment is started during the first 30 to 50 days of life.

If the tilapia are reared in cages they cannot breed successfully as there is no pond bottom on which to spawn, or if they do spawn the eggs fall out of the cage and are lost. The disadvantage of this system is that the fish have to be fed by the farmer.

Some species can grow in salt and in brackish water, for example *S. mossambica* and *S. aurea*. However, they cannot breed under these saline conditions and so overcrowding does not occur.

15.1   Egg production

Egg production presents no problem as the fish readily spawn in the ponds.

A typical spawning pond is $100\,m^2$ and 1 m deep, stocked at density of 12 females to four males. Such a pond will produce a steady supply of 2000–5000 fry every 3–4 months. During the early stages the fry feed on the natural food produced by fertilization in the pond. The fry are removed from the fry pond to the growout pond, where one of the previously described systems for preventing over-population is used.

15.2   Growout ponds

The size of these ponds depends on a variety of factors and is not critical. The size is decided by the topography of the ground, the amount of fertilizers available to the farmer and the size of the harvest desired. It varies from a few hundred square metres to several hectares. Typical intensive cultivation units are about 800–1000 m². These ponds are a practical size for the farmer to manage.

15.3   Feed

The young stages do not take any artificial feed but rely entirely on the natural production of the pond. Adult tilapia can be raised solely on the natural production in the pond resulting from manuring and/or fertilizing with artificial fertilizers. This natural feed production can be supplemented, to a greater or lesser extent, by the addition of other food stuffs. Tilapia can be fed a variety of feeds including plant leaves, suitable for herbivorous species of tilapia, and meal and various waste products suitable for other species. The main plants used include cassava, sweet potato, cane, maize and pawpaw. Waste products include rice bran, waste cotton seed cake and peanut cake, fruit, brewery waste and coffee pulp, the type of food used depending on the availability and local cost.

The efficiency with which the fish can convert these various types of feed varies considerably with the feed, the conversion ratio varying from about 1:60 for leaves to 1:3 for various oil cakes.

15.4   Fertilization

Inorganic fertilizers

In many countries the use of inorganic fertilizers for the production of tilapia is uneconomic because the fertilizer is expensive. However, their use is possible in intensive production systems. The amount used is adjusted according to the pond requirement but a typical example would be the use of double superphosphate fertilizer at the rate of 700 kg/ha/year.

Organic fertilizers

Most tilapia farms are based on the use of organic fertilizers. These are applied either at intervals to the pond, or else, in the case of pig and poultry manure, continuously as the animals are farmed in conjunction with the fish ponds. Ducks are kept on the fish pond so that their droppings are released directly into the water, and pigs are kept alongside the ponds so that their manure can be washed directly into the pond.

The amount of fertilizer required varies with the pond fertility and the stocking density. Typical application rates are:

Pig manure   10 000–30 000 kg/ha/year or two animals/
                   100 m^2
Poultry   10 000–12 000 kg/ha/year or 10–50 ducks/
                   100 m^2
Cattle   30 000 kg/ha/year

15.5 Stocking rate

The stocking rate is determined in the light of the production system used, the amount of fertilizer and supplementary food available and the number of fish required. In general, for the straight production of tilapia a stocking rate of 10 000–30 000/ ha can be used. Stocking rates for production systems using a predator are generally 10 000 tilapia/ha the ratio of predators varying with the species of predator used; for example *L. niloticus* is stocked at 50/ha, tucunare at a level of 10% and *H. fasciatus* at a level of 2–5%.

15.6 Production levels

Obviously the production level possible varies with the type of farming practised; typical levels for the different systems are:

Unfertilized ponds with no predator	300–600 kg/ha/yr
Ponds fertilized with pig manure	5000 kg/ha/yr
Ponds fertilized with poultry manure	3000 kg/ha/yr
Ponds fertilized and supplementary feed given	6000 kg/ha/yr
Ponds fertilized and supplementary feed given and all-male populations farmed	9000 kg/ha/yr

Chapter 16
Carp

Carp belong to the freshwater family Cyprinidae. This is a wide-spread and abundant family absent only from South America, Madagascar and Australia in their natural distribution. The relationships between the species in the family are poorly understood, the family comprising some 1600 different species. Of these only very few are of much importance in fish farming. Farmed carp are divided into three groups:

Common carp, which is farmed extensively in Europe, Asia and the Far East.
Indian major carp, so called to exclude numerous other less important carp species included in Indian fish farms.
Chinese carp.

These groupings contain the following species:

Common carp:
Cyprinus carpio	carp	omnivorous feeder

Indian carp:
Catla catla	catla	surface feeder on plankton and vegetable debris
Labeo rohita	rohu	midwater feeder on vegetable debris
Cirrhinas mrigala	mrigall	bottom feeder

Chinese carp:
Ctenopharyngodon idellas	grass carp	higher plants
Hypophthalmichthys molitrix	silver carp	phytoplankton
Aristichthys nobilis	big head	zooplankton
Mylopharyngodon piceus	black carp	molluscs
Cirrhinas molitorella	mud carp	detritus

As can be seen from this list the carp contain species which feed on different foodstuffs. This fact has been utilized by Chinese fish farmers to develop a system of polyculture of carp

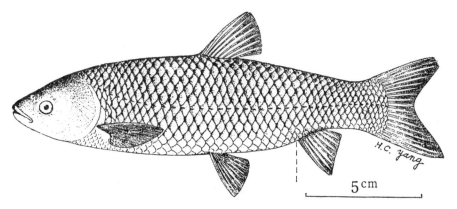

Fig. 16.1 Common carp (*Cyprinus carpio*) (from Chen).

in which those species feeding on various parts of the food chain in a pond are kept together in the same pond. They therefore do not compete directly for the available feed and in this way a much higher production of fish can be obtained than would be possible with the production of a single species.

16.1   Common carp

This is a freshwater fish reaching 80 cm and 10–15 kg. The temperature range is from 1 to 40 °C; the fish grows at temperatures above 13 °C and reproduces at temperatures above 20 °C. It is mature after 3 years and spawns each year in spring in temperate climates; in the tropics it will spawn every 3 months. The female fish spawn 100 000 eggs/kg body weight.

Growth is rapid in the tropics, where the fish can reach a weight of 400–500 g in 6 months. They are omnivorous.

Egg production

Egg production presents no problems with these fish as they will spawn readily in farm ponds. Various systems are used differing in the degree of control that the farmer can exercise over the procedure.

In the simplest system the fish are allowed to spawn in special ponds and the brood fish are removed after spawning. The ponds are $8-10\,m^2$ in size and 20 cm deep in a central region which is covered with grass. Either one end of the pond or a peripheral ditch is made 50 cm deep. The ponds are stocked with one to three sets of fish, each set consisting of two females and one male. Generally 3-year-old males and 4-year-old females are used. The eggs are small, $1-15\,mm$ in diameter; they are sticky and adhere to the grass in the centre of the pond where spawning occurs.

The eggs hatch in about 4 days and the fry are about 5 mm in length.

There are various variations on this basic system which are mainly concerned with different techniques for collecting the eggs from the spawning pond. In some systems conifer branches are placed in the pond and are removed after spawning with the eggs attached to them and placed in a hatching pond. Another variation on this is to place floating plants in the pond to act as egg collectors. These plants are often used in conjunction with cloth tanks, forming a spawning region inside a main pond.

In Indonesia fibre mats called 'kakabans' made from palm fronds are used as egg collectors. Kakabans are used in ponds 24 to $30\,m^2$ in area and five to seven mats each 1.5 by 0.5 m are placed in each pond. The ponds are stocked at a rate of 8 kg of female fish/$5\,m^2$ of pond area. After spawning the mats are removed to hatching ponds. These are generally about $600\,m^2$ in area and each set of mats is suspended some 10 cm below the water surface.

The technique of induced spawning is also used in carp production. It is often used in temperate water farms, where spawning of the fish is less predictable than in tropical farms.

Ripe fish are injected with carp pituitaries at the rate of one pituitary per kg of body weight. The pituitary is macerated in saline solution, 6 g of sodium chloride/litre of water. Dried pituitaries are also used in the ratio of $2-3\,mg$ of acetone-dried pituitaries per kg of body weight of fish. The fish are hand stripped $12-20$ hours after injection. The mixture of eggs and milt is then added to what is known as fertilization fluid. This comprises 10 litres of water with 30 g urea and 40 g of sodium chloride dissolved in it. The fertilized eggs are sticky and so, to be able to hatch them in hatchery jars, the eggs are treated with a solution of 15 g of tannin in 10 litres of water to remove the stickiness.

The eggs are incubated in incubation jars, 120 000 eggs requiring a water flow of 2 litres/min. The eggs hatch in 4–5 days.

Nursery ponds

These are 0.25–2 ha in area depending on the size of the farm. The ponds are 0.5–1.5 m deep. They are stocked at a density determined by the water flow possible into the pond. In still water 50 000/ha, in running water 300 000–800 000/ha.

Growout ponds

The type of growout system required for carp depends on climatic conditions and market requirements. In Europe, where water temperatures are comparatively low and large fish are required for the market, 3 years are commonly required to produce market-size fish. In Israel market-size fish of between 500 and 700 g are produced in 1 year and in tropical countries 500 g fish can be produced in 6 months and 1–1.5 kg fish in one year.

In Europe three stages of pond production are required. In the first stage 4–8-week-old fry are stocked in ponds 0.25–2 ha in area and 70 cm deep. Natural food production is stimulated by fertilization and the fry are stocked at a density of 50 000/ha. The second stage is in ponds 2–25 ha in area, depending on the size of the farm, and 1.5 to 2 m deep. The ponds are sown with lupins and grass before filling and are stocked at 5000/ha. The fish over-winter in the ponds and 9–12 cm fingerlings are produced. The final stage ponds are 2 m deep and 10 ha to hundreds of hectares in area, the size of the ponds being decided by the topography of the ground, the needs of the farm and the availability of fertilizer. The method of treating the ponds varies, again according to circumstances. Sometimes the ponds are untreated; some farms use animal fertilizer, others use treated sewage; some farm in conjunction with ducks or pigs, and other farms use supplementary feeds, generally vegetable waste materials.

Production

Production levels achieved vary with the intensity of the farming, the level of fertilization and also with the water temperature. Some examples are as follows:

Untreated ponds	25–400 kg/ha
Fertilized ponds	2000 kg/ha
Supplementary feed	2000 kg/ha
Treated sewage	900 kg/ha
Fertilized by ducks	500 kg/ha

In the tropics the production level varies from 300 kg/ha in unfed unfertilized ponds up to a claimed 80 000 kg/ha in intensive feeding in running water in the Philippines.

16.2   Indian and Chinese carp

These freshwater fish cannot withstand low water temperatures and have an optimum growth at about 25 °C. They are sexually mature at between 4 and 9 years and will spawn only in temperatures above 25 °C. The age of sexual maturation depends on their growth rate, which in turn reflects the number of months spent in water temperatures above 20 °C. Mature fish weigh 5 kg or more. These fish, unlike the majority of *Cyprinidae* species, lay pelagic eggs which float in the water of rivers during the incubation period.

Egg production

Until recently the supply of seed for this industry has depended totally on the collection of eggs, fry and fingerlings from the rivers in which the adults spawn. The collection, distribution and sale of this seed is on a commercial basis, with dealers selling to fish farmers often with farms a considerable distance from the seed collection point. The eggs and fry are caught in fine mesh nets generally fixed in the water in wooden frames. The eggs caught in this way are hatched in fine netting cages near the collection area; the eggs hatch in 2–3 days. The fry are sorted by experts into species before being transported for sale.

There was a great increase in the production of seed by induced spawning after this technique had been developed for these species some years ago. Spawning fish, or fish which are nearly ripe, are selected from the farm stock and injected with acetone-dried fish pituitaries, or with the hormone chorionic gonadotropin. The fish receive 2–3 mg of dried pituitaries/kg of body weight. Both males and females are treated and the females receive a second dose of about 3 mg/kg. If hormone is used the dosage level is 700–1000 International Units/kg of body weight. Grass carp require a slightly higher dosage of fish pituitaries.

After injection the fish are placed either in spawning ponds which are 100 m² and 2 m deep or else in special netting cages suspended in the pond water. These cages are 6 m² and 1 m deep. Generally two males are placed together with one female. Spawning occurs in about 12 hours.

The eggs are collected by hand net and transferred to the hatchery facility, generally indoors. The eggs are placed in trays in running water or in nylon baskets in running water or in hatchery jars. In jars the egg density is 5000–10 000/litre with a water flow of 0.7 litre/min. In trays or boxes, which are housed in troughs, water flow rates of 20 litre/min are used. The eggs hatch in 24–30 hours, depending on temperature; 24 hours are needed at 28 °C and 30 hours at 25 °C.

The larvae have small yolk sacs which are absorbed in 3–6 days. The fry are then ready for transfer to nursery ponds.

Nursery ponds

The area of these facilities varies considerably from country to country. In India, for example, small ponds of 10 m² are sometimes used, whereas in China pond size varies up to 2 ha in area. The ponds are shallow, 0.5–1 m in depth. Inside these ponds the fry are sometimes kept in floating mesh cages before release into the pond proper. The ponds are stocked at 200 000/ha.

Before stocking, the ponds are prepared by the application of manure at a rate of 2000 kg/ha, to stimulate plankton growth.

Supplementary feeding is sometimes used but the principal feed of the fry is plankton and any supplementary feed probably acts mainly as an additional fertilizer.

Growout ponds

These ponds vary in size from 1 ha to a few hectares. In China the ponds are often 2–3 m deep. These ponds are stocked at densities of about 6000–10 000 fry/ha, the proportion of fry of different species varying to suit local conditions and the need of the farmer. The fish may be left 3 years before harvesting. The average yield is about 4000 kg/ha but higher yields are often obtained.

Indian carp are grown in smaller, shallower ponds about 0.5 m deep. As with Chinese carp, the Indian carp are stocked in ratios of different species to suit local conditions and to meet local needs. Production of Indian carp is about 900 kg/ha, which is harvested as 300 g fish in 8 months.

The growth of both types is sustained by a programme of fertilization. All types of manure are used and supplementary feeds, mainly of plant materials, are given. Snails are sometimes given in farms producing mollusc-eating carp. Grass carp are excellent species for controlling weed growth in ponds so naturally growing aquatic plants, such as the macrophytes, can form a significant part of the diet. Often pigs and ducks are kept in conjunction with the carp farms, 100 pigs or 2500 ducks/ha being used.

Chapter 17
Milkfish

The milkfish, *Chanos chanos*, is a semi-tropical and tropical marine fish widely spread throughout the Indian and Pacific oceans. The fish grows to a length of 0.75–1.5 m and up to 18 kg in weight. It has a temperature range of between 15° to 40 °C and can tolerate low oxygen levels. The salinity range of the fish is extensive, stretching from nearly fresh water to full sea water and, in the special case of fish living in lagoons on Christmas Island, up to 70‰ salinity can be tolerated.

The fish spawn in certain distinct regions of the oceans over about 30 fathoms of water depth. Each female lays about 5 million eggs which are planktonic and hatch in 24 hours. The larvae are pelagic and drift in the ocean currents, which carry them into the offshore waters in various regions. The larvae migrate into the coastal waters and have reached the fry stage in their development when they arrive there. The fry migrate into the shallow water of coastal lagoons and river mouths. They appear in coastal waters in two seasons of the year, March to June and August to September, depending on the geographical location. The peak occurrence is in May.

The young stay inshore for about 1 year, by which time they are about 20 cm long and weigh 200 g. After this they migrate to sea. The adult fish feed on plankton in the wild and, under farm conditions, on the algal mat growing on the pond bottom. They become sexually mature at about six years old.

17.1 Egg production

Egg production is still not possible in commercial quantities on farms, although an active development programme is in progress to develop the technology for induced spawning of the

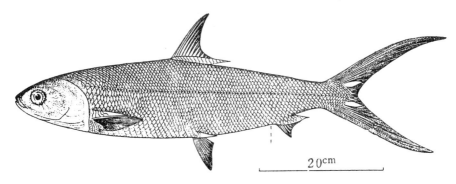

Fig. 17.1 Milkfish (*Chanos chanos*) (from Chen).

adults and the rearing of the fry. At present farming depends on the collection of naturally-produced fry. The fry collection is carried out in coastal waters, either by farmers or by professional catchers who sell direct to the farmers or via middle men. Hand push nets are commonly used as the fry tend to concentrate in tidal creeks and in pools left behind after high tide. The occurrence of the fry is influenced by lunar and tidal periodicity, which is superimposed on the seasonal variation in fry numbers. Fry numbers vary annually from country to country and constitute a hazardous, uncontrollable factor in the development of milkfish farming.

17.2   Nursery ponds

The ponds used for the rearing of fry and fingerlings are shallow, with a depth of no more than 10 cm. The area varies between 500 and 5000 m², depending on local conditions and the requirements of the farmer. The bottoms of the ponds are ploughed when they are dry and limed if necessary to control the acidity. They are filled with dilute sea water, at 10–15‰ salinity, before the introduction of the fry. The ponds are fertilized with manure at a rate of 400–1000 kg/ha. Rice bran is also used as a fertilizer. Chicken manure at a rate of 2000 kg/ha is used in some countries where available. The objective of this treatment is to establish a layer of green algae and associated diatoms and animals which, together with the plankton, forms the food of the animals. For the early fry stages the plankton is a more important element in the food than is the bottom layer called 'lab-lab'.

Depending on local conditions the fry ponds are stocked at between 70 000 and 500 000 fry/ha. The fry are transferred to growout ponds when they are between 2 to 10 g in weight.

17.3   Growout ponds

These ponds vary in area between 0.5 ha and many hectares, depending on local conditions. They are constructed so that they are connected to the sea via a sluice gate which allows the pond water to be part-exchanged with each tide cycle. This refreshment of the water allows the required level of oxygen and nutrients to be maintained in the pond. Extreme variations in the water level are prevented by a system of baffle boards at the sluice gates, which are screened to prevent the entrance of predators and the exit of young milkfish.

The ponds are shallow, about 70 cm in depth, and the bottoms are fertilized when the pond is empty. Many different substances are used as fertilizers, ranging from manures to plant wastes including green manure, rice bran, peanut cake, soya bean cake, and chicken manure. The objective of this manuring programme is to build up a layer of 'lab-lab' some 0.5 cm deep over the pond bottom. After manuring, the pond is covered with a shallow layer of sea water which is allowed to evaporate. The pond is then refilled and at this stage undesirable fish and insect species are eradicated by the use of chemicals or tea seed meal.

The ponds are stocked at between 2000 and 10 000 fish/ha. During the growing period more fertilizers and supplementary feeds are added to maintain the layer of 'lab-lab'. Such things as rice bran, soya bean meal, peanut meal and copra meal can be added at a rate of 25 kg/ha/day.

The growout ponds are stocked with fish when the fish have reached 2–10 g weight.

17.4   Productivity

This varies with the size of the fish harvested, the fertility of the pond and the use of supplementary feeding and fertilization.

Production by traditional methods, which relies on low levels of fertilization and often uncontrolled stocking when the fry enter with the tidal water (which also brings in predators), is low at about 800 kg/ha/year. The use of supplementary feeds and an improved programme of fertilization raise these yields to 3000 kg/ha/year. The size of the harvested fish and the rate of growth varies according to farming technique and country. For example in the Philippines farmers produce 450 g fish in 9 months and in Taiwan 250–600 g fish in 4 months.

A development programme has demonstrated that milk-fish of about 5–7 g weight can be used as baitfish in the tuna fishing industry. These fish can be produced by normal farming procedures and supplementary feeding allows production levels of 2800 kg/ha/year to be achieved.

Chapter 18
Mullet

There are many different species of mullet which all belong to the family *Mugilidae*. These fish are widely spread throughout the tropical and temperate seas ranging from latitude 42° north to latitude 42° south. They form an important coastal fishery in many parts of the world. Because they are widespread and because the differences between the species are often slight, much confusion has arisen over systematic classification. At present there are recognized about 100 different species and of these the following have been reported as being cultivated:

Mugil cephalus   Mediterranean region
M. chelo
M. capito
M. saliens
M. oligolepis

M. cephalus   Black Sea and Caspian Sea regions
M. saliens
M. auratus

M. cephalus   Southeast Asia and Far East
M. dussumieri
M. troschelli
M. wagiensis
M. seheli
M. macrolepis
M. corsula
M. tade
M. engeli
M. parsia

M. falcipinnus   Nigeria
M. grandisquamis

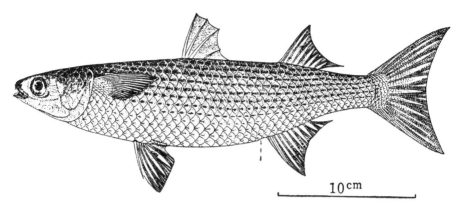

Fig. 18.1 Grey mullet (*Mugil cephalus*) (from Chen).

Of these species by far the most important is *M. cephalus.*

The mullet inhabit coastal sea water but migrate into brackish water and even fresh water as they have a wide range of salinity tolerance. Their temperature range is from 3 to 30 °C.

The mullet is a detritus filter feeder. It sucks mud from the pond bottom under farm conditions and filters out its food by the use of filters located in the mouth and on the gill arches. It will also feed on the surface scum of the pond. In the sea the mullet feed on planktonic algae and benthic algae. Mullet reach maturity at a fairly small size, 30 cm long and 300 g weight, but they can reach 4 kg weight.

18.1   Egg production

Mullet spawn in full sea water, usually at the edge of the continental shelf over deep water. Each female lays a large number of very small eggs which, after fertilization, float in the plankton and hatch in about 2 days. The number of eggs produced varies with the species, *M. cephalus* for example laying between one million and four million eggs. Upon hatching, the larvae, which feed on the plankton, drift inshore, arriving there in large numbers when they are about 25 mm in length. In inshore waters the larvae form large shoals which migrate along the shore and enter freshwater areas. At this stage the larvae are caught by fish farmers using hand nets and seines with a fine, usually 6 mm, mesh.

Because of the great importance of mullet as farm fish considerable research and development work has been undertaken to develop the technology for the induced spawning of farm fish and the rearing of the resultant eggs and larvae. This technology now exists but as yet has not been extensively applied. The induced spawning system depends on the use of fish pituitaries and pituitary hormones as in the other induced spawning techniques that have been developed in aquaculture. This technology is very fully detailed in the review by Nash and Shehadeh (see bibliography).

18.2   Nursery ponds

Fry from the wild are reared in shallow ponds 0.5 m deep, the area of which varies with the requirement of the individual farm, usually between 2 and 4 ha. The young fish feed on the plankton and so it is necessary to fertilize the ponds before the introduction of the fish and during their growth in order to stimulate and maintain the required level of plankton production. Much use is made of fertilizers, although some manuring is also practised. Levels of fertilizer used are about 200 kg/ha of superphosphate applied once a year and 300 kg/ha of ammonium sulphate applied once a week.

The ponds are stocked at a density of between 10 000 and 20 000 fry/ha.

In the hatchery system the fry are fed with *Artemia* larvae and with rotifers.

The young stages are very sensitive to extremes of environmental change, so that care has to be taken when the wild fry are first introduced into the rearing ponds, where high mortalities can be caused by great changes in salinity and water temperature.

The fry are kept in these ponds for about 2 months, during which time mortalities as high as 20–30% are common.

18.3   Growout ponds

These are generally 1 m deep and some 4–6 ha in area. As with the fry ponds these are fertilized by the use of artificial fertilizers and also manure. Artificial fertilizers are used at the same rate as in the fry ponds. The growing ponds are stocked at a density of around 5000/ha.

Supplementary feeds are used which are eaten directly by the fish but also act as fertilizers to develop the fauna of the bottom mud of the pond on which the fish feed. Rice bran, peanut cake and soya bean cake are all used as feed. The total feed given varies with local conditions and the productivity of the pond; levels of 5000 kg/ha/year are common.

Under tropical conditions *M. cephalus* can reach a weight of 1.5 kg in 2 years and 300 g in the first year. Production levels vary with the intensity of the farming practised; with supplementary feeds levels of up to 2500 kg/ha/year can be achieved but unsophisticated farming procedures will yield levels as low as 300 kg/ha/year.

Mullet are very often used in mixed culture systems with either carp or tilapia. In this way the maximum value of the detritus feeding habit of the mullet can be obtained as the mullet will not compete for food with the other fish.

Chapter 19
Turbot

The turbot (*Scophthalmus maximus*) is a large marine flat fish much esteemed as food and commanding a high market price. It is the latest, and potentially the most important, marine fish to be included in farming in temperate seas. For this reason it was felt that a brief introduction to the farming of this species should be included in the second edition, even though, at this date, the system is only just past the development stage.

Work on the farming of marine fish in temperate seas was started over 40 years ago with research work on the farming of plaice (*Pleuronectes platessa*). This work established the feasibility of farming these types of fish, but it had no commercial future because of the low market value of the product. Recently turbot have been successfully bred and reared under farm conditions, thus establishing the beginning of what will undoubtedly be an important marine farming system in the future. Juvenile production of turbot is taking place mostly in N.W. Europe whilst on-growing is concentrated in the warmer waters of southern Europe, especially around Spain and

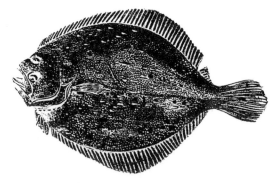

Fig. 19.1 Turbot (*Scophthalmus maximus*).

Portugal. The fish are docile and grow well under farm conditions; they spawn readily and their breeding season can be extended by manipulation of the day-length. Production fish are fed pelleted rations. The success of this work has already stimulated research work on other species of flat fish, especially halibut. The next five years should see a major development of this industry.

Chapter 20
Eels

Eels are cultivated extensively in eastern Asia and to a limited extent in Europe. Eels are catadromous, that is to say, they spawn in the sea and migrate when very young into fresh water where they grow to maturity. Both European and American eels spawn in the Sargasso Sea in the Atlantic Ocean. The larvae drift in the ocean currents and, in the case of the European eel, they are carried in the Gulf Stream across the Atlantic to Europe. During this voyage they change from leaf-like larvae to very small eels called elvers, which ascend the rivers where they grow to maturity. When mature they return to the sea.

Three species of eel are of economic importance:

European eel   *Anguilla vulgaris*
American eel   *Anguilla rostrata*
Japanese eel   *Anguilla japonica*

European eels spend up to 3 years at sea before entering fresh water; the other two species spend only one year at sea.

Eels have not as yet been reared from artificially fertilized eggs and so the farming of these species is totally dependent on the collection of wild seed, elvers.

20.1   Extensive production

This type of farming is practised in Italy, where the young eels enter coastal brackish water ponds and lagoons in the course of their natural migration. The adult eels are caught in fixed traps when they migrate to sea. Production varies between 30 kg and 90 kg/ha.

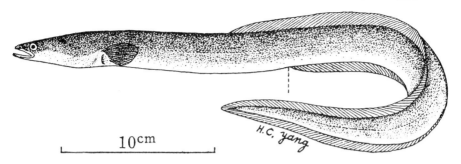

Fig. 20.1 Japanese eel (*Anguilla japonica*) (from Chen).

20.2   Intensive production

In eastern Asia intensive production of eels has been practised for many years and current production in Japan exceeds 30 000 t per year. Eels are also farmed in Taiwan, China, South Korea, Thailand and more recently in Malaysia.

20.3   Nursery ponds

These are 100–300 m² in size with a depth of 40 cm. In Japan ponds are of earth and in Taiwan mainly brick and concrete. Elvers are captured in winter when they are 6–7 cm in length. They are stocked at a rate of 300 000 per ha. During the course of the first year the eels are graded five or six times. After each grading ponds are restocked with fish of similar size. This is done to reduce the incidence of cannibalism, which is a serious problem in eel farming.

20.4   Growout ponds

These are 4000–10 000 m² in size with a depth of 1 to 1.5 m. The ponds are supplied with running fresh water, which is often aerated to maintain the required level of oxygen. The ponds are stocked at a density of 1.2 to 2 kg/m². The growing period is between April and November, when the water temperature exceeds 15 °C. Production levels of 10–20 t/ha are obtained.

The fish are fed on chopped sea fish, mussels, silk worm cocoons and *Tubifex* worms. Dried powdered food has now been introduced which is specially formulated for the eel. The feed is presented in boxes covered with a wire mesh which are suspended in the water. Extruded pellets are also gaining inpopularity since they have a better uptake and are less likely to reduce water quality.

High production losses of 10–40% are sustained during the production cycle. The high local prices which are obtained for these fish must be an important factor in the economy of this type of farming.

Intensive eel farming has been tried at an experimental level in Germany. The fish were stocked at a level of $1.5 \, kg/m^2$ with elvers weighing between 3 and 35 g. The fish were fed chopped fresh fish and dried food. At maturity the males weighed 90 g average and the females 500–600 g. Food was given at a rate of 10% of body weight per day.

Chapter 21
Marine Shrimp

Marine shrimp farmed in different parts of the world belong to two families, the Penaeidae and the Crangonidae. In all some 21 different species are produced in farms of greatly varying levels of development. The species are distributed as follows:

Penaeus indicus	Indian shrimp	Asia
P. merguiensis	banana shrimp	
P. monodon	black tiger shrimp	
P. semisulcatus	green tiger shrimp	
P. carinatus		
P. teraoi		
P. japonicus	kuruma shrimp	
P. orientalis		
P. chinensis	Chinese white shrimp	
Metapenaeus brevicornis	yellow shrimp	
M. burkenroadi		
M. ensis	sand shrimp	
M. dobsoni		
P. occidentalis		Central America
P. stylirostris	blue prawn	Pacific Coast
P. californiensis	brown prawn	
P. vannemei	white prawn	
P. aztecus	brown shrimp	Gulf of Mexico
P. setiferus	white shrimp	(little if any farmed)
P. duorarum	pink shrimp	
P. brasiliensis	Caribbean brown shrimp	

The marine shrimp is distributed throughout the oceans of the world but the commercially important species listed above inhabit the shallower parts. They are animals capable of fast growth and, depending on species, may reach an adult size of 18–30 cm in length. The females are fertilized by the males during their intermoult period and the eggs are laid soon after

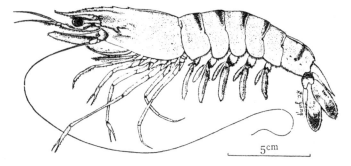

Fig. 21.1 Grass shrimp (*Penaeus monodon*) (from Chen).

the hardening of the female exoskeleton. The eggs are released into the sea, where they hatch and complete a complicated life cycle during which they pass through different stages, each with a different body form, nauplei, zoea, mysis and postlarval stages. The postlarval stage has the same body shape as the adult although, of course, of a much smaller size. The postlarvae migrate into inshore brackish waters where there are rich feeding grounds in the shallow areas of the coast.

Many different species of shrimp are farmed, the procedures used varying from simple entrapment of mixed species to the production of the eggs and young inside the farm system.

Shrimp are omnivorous detritus eaters which in farm ponds live on the algal mat growing in the pond or on animal and plant food added as food. It is probable that a high proportion of the added food acts as a fertilizer and not as a direct food for the shrimp.

21.1   Egg production

Shrimp will not spawn directly in farm ponds and the complicated larval development cannot take place there. The seed for the farm is therefore either obtained from the wild from the migrating stocks of young wild shrimp or else produced in the farm either from wild fertile females or from females reared and mated in the farm.

Wild seed

The wild seed are caught by the farmers in the inshore waters by means of hand nets. In some farms where the production systems are uncontrolled, the seed are allowed to enter the ponds with the tidal entry of the water. This latter system leads to the farming of mixed species and the introduction of predators and competitors, which greatly reduces production levels.

Use of wild females

The females are fertilized in the sea before capture and are selected by inspection to ensure that they have ripe ovaries. These females are then taken to the farm, where they will spawn in the spawning ponds. Often spawning is hastened and made more certain by the removal of one eye stalk or the contents of one eye stalk. The ablation of the stalk, as it is called, removes part of an important endocrine organ of the shrimp and thus upsets the hormonal balance, bringing about spawning of the female. The females are stocked in concrete tanks lined with epoxy resin at a density of two per m^3 of water. Each female spawns about 200 000 eggs. Probably only half of the females

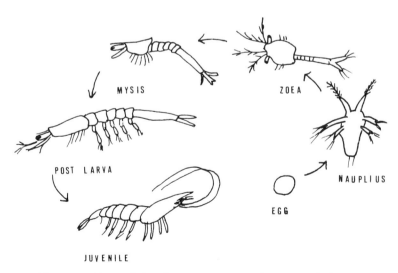

Fig. 21.2 Typical development stages of a penaeid shrimp.

will spawn and only half of the eggs will hatch. The eggs hatch in 14 hours at 28 °C. The tanks are aerated with oil-free air.

The eggs hatch into nauplii, which carry a yolk sac and do not feed. The nauplii undergo six moults during 36 hours before changing into the next stage called zoea. The zoea moult three times before they change into mysis. This takes about 4 days. During this time the zoea are fed on diatoms, chiefly *Skeletonema costatum*, which are produced in separate culture tanks. The density of this feed is maintained at 3000–10 000 skeletonema cells/cm^3 of larval tank water. Sometimes the water in the tank is fertilized in order to stimulate the production of the diatoms. However, this procedure reduces the control of the larval tank as over-production of the diatoms may occur.

The mysis moult three times in 3 days before changing into miniature adult forms called postlarvae, known as PLs. The mysis are fed on diatoms and on the nauplii of the brine shrimp, *Artemia salina*. *Artemia* lives in various parts of the world in salt lakes which dry out at certain times of the year. When this happens the *Artemia* produces a very resistant cyst inside which the resting stage of the animal can live for long periods, hatching out when conditions again become favourable for its growth. These cysts are collected and packed under vacuum and sold to the aquaculture industry. They are used in shrimp farming and in the production of fish which have very small larval stages. When placed in water at the correct temperature and salinity the cysts hatch and produce free-swimming larvae of the brine shrimp, which serve as a good food for the cultured shrimp. In shrimp farming the cysts are either placed directly into the shrimp tanks where they hatch (about 5 g of cysts per day for every 10 000 mysis) or the *Artemia* are hatched in separate tanks. Rotifers, small aquatic organisms, are also cultured in separate tanks and fed to the shrimp. The postlarvae are fed on *Artemia* nauplii for about 10 days and then they are fed a variety of feed such as minced clam, egg yolks and manufactured dry feed. Feeding levels of wet feed are 2–3 times the body weight of the postlarvae per day.

During the postlarval stage the depth of sea water in the tanks, which starts at about 80 cm, is increased by 15 cm per day until a depth of 2 m is reached; thereafter the water is exchanged at a rate of 20–40% per day. After about 30 days the postlarvae should have reached an average weight of 20 mg, at which stage they are ready for transfer to the nursery ponds.

Use of females mated in the farm

This system is as yet confined to use in only a few advanced hatcheries in the world. Considerable investment has been put into the development of these processes and details are slowly becoming available. The object of the system is to obtain mature males and females from the farm stock which will mate under farm conditions. After mating the ripe females are treated in the same way as are ripe females taken from the wild. The secret of the procedure lies in obtaining healthy stock and keeping them under the correct environmental conditions and feeding them with the correct food. In general the following factors appear to be important:

Brood stock grown from postlarvae to maturity in 10 months.
Males of over 50 g weight and females over 70 g weight.
Females have one eye stalk removed.
More males than females are kept together.
Tanks must be at least 3 m^3 of water.
The pH must be between 8.0 and 8.2.
Salinity must be above 30‰.
The shrimp must be fed fresh seafood.
Certain levels of illumination are required.

21.2   Nursery ponds

The areas of these ponds are about 0.1 ha with a water depth of about 80 cm. The postlarvae are stocked at a density of one million per ha. The survival in the nursery ponds is about 70% and the postlarvae should reach an average weight of about 1 g after 50 days. The postlarvae are fed a variety of food such as minced clam or animal or plant waste or manufactured feed. As the exact dietary requirements of the shrimp is not as well understood as it is for salmonid fish, the formulation of these feeds cannot be as exact. Often, chicken pellets are used for the shrimp feed. It is probable that much of the feed is not consumed directly by the shrimp but acts as a fertilizer stimulating the growth of the bottom mat of algae on which the shrimp graze.

The feeding rate varies with the stage of development of the shrimp and is related to their body weight. For the first 7 days 200% of the body weight is fed per day, then 50% for the next 14 days and 25% for the last 21 days in the nursery ponds.

21.3   Growout ponds

The size of these ponds varies considerably from well below 1 ha to many hectares in area. Many farmers consider that an area of about 2 ha is the best size; such ponds have a sloping bottom and an average depth of 1 m. The shrimp from the nursery ponds are stocked at a density of 100 000 per ha.

Depending on water temperature and feed, the shrimp can reach a marketable weight of 25 g in 4–6 months. The survival during this period is about 90%, the actual figure depending on the skill of the farmer.

The shrimp are fed a variety of feed, depending on local availability and the sophistication of the farming procedures used. These feeds include clam meat, animal and plant wastes, formulated pellets and chicken pellets. Again, as in the nursery ponds, a significant amount of the food is probably acting as a fertilizer for the production of the algal mat. When clam meat is used as a feed an overall conversion of 10:1 is often achieved; with pellets, conversion ratios of 4:1 and sometimes 2:1 are obtained.

21.4   Pond management

Shrimp farming in many areas of the world has recently led to severe problems of environmental degradation. The control of water quality is therefore a key factor for regulating success in shrimp farming.

Salinity, temperature, oxygen level and hydrogen sulphide level are the most important parameters for the management of the pond water. Regulation of these factors is mainly achieved by control of water exchange, either by the use of tidal flow or by pumping. Oxygen levels are often maintained by the use of mechanical agitators to aerate the water. Hydrogen sulphide levels are controlled by the use of iron oxide which is added at the rate of 1 kg/m^2 of pond bottom per day. Water exchange is, however, of the utmost importance, and intensive feeding systems require an exchange rate of 10%/day which can be reduced to 5% in less intensive systems. It has been calculated that, if water exchange rates of 33%/day are possible, very high production levels for shrimp, in the order of 5000 kg/ha/year, can be achieved.

21.5 Production levels

These vary enormously from farm to farm depending on the farm management and the type of farming practice, the levels varying from a few hundred kg up to in excess of 3000 kg/ha/ year.

Harvesting can often take place about 150 days after stocking when the shrimp sizes will be about 22 g weight.

Shrimp are frequently marketed by their size, calculated as the number per pound (lb), either whole or as tails. In intensive systems counts of 30–40 per pound of headless shrimp can be obtained.

Chapter 22
Freshwater Prawns

'Freshwater prawns' is a term used to decribe prawns which spend their whole life in fresh water as well as those that only spend part of their life cycle in fresh water and the rest in sea water or brackish water. At present 19 different species of freshwater prawns have been adapted to farming conditions either on an experimental basis or at an industrial level.

All belong to the genera *Leander*, *Palaemon* and *Macrobrachium*. The vast majority, however, belong to *Macrobrachium*. Of these by far the most important is the so-called giant freshwater prawn of the Indo-Pacific region *Macrobrachium rosenbergi*. This prawn, which grows to a length of 30 cm, excluding the length of the front claws, is now successfully farmed in many countries. The adult prawns live in fresh water but the larval stages require brackish water for their development. The adult prawns are omnivorous animals which feed on the food they find in the bottom debris of a pond.

22.1 Egg production

In freshwater prawns, egg production presents fewer problems than in marine shrimp, as the adults become sexually mature and mate under farm conditions. The females are fertilized by the males during the times between moults when their shell is soft. Egg laying occurs when the shell has hardened and a few hours after mating. The eggs are carried by the female under the abdomen until they are hatched. The female lays about 100 000 eggs which take about 20 days to hatch at 28 °C. After spawning the female is ripe again after about 23 days. The eggs hatch into larvae which swim freely in the water, where they go through a number of developmental stages which are completed in 2–3 weeks. The larvae feed on zooplankton during their development. When they have metamorphosed into the adult form the young prawns migrate back into fresh water for the adult stage of their life.

Ripe fertilized females for egg production are obtained from the farm ponds and, where possible, from the wild. The females are placed in indoor concrete tanks lined with epoxy resin. The water depth is about 70 cm with a salinity of 12‰ and a temperature of 26–31 °C. The water is exchanged at a rate of 50%/day. Three females are stocked for every m^3 of water. This stocking density of females should produce a larval density of 50 per litre of water. Generally the tanks are exposed to subdued light with about 90% of the surface covered. The eggs, incubated on the underside of the female's abdomen, hatch under farm conditions in about 20 days at 28 °C. The larvae go through various moults and change into the adult form postlarvae in 3 to several weeks, depending on food and conditions.

The larvae are fed *Artemia* nauplii for the first 10 days after which time they are fed *Artemia* nauplii and shredded fish flesh. The feeding rate is 30 g of *Artemia* cysts and 90 g of fish flesh/1000 postlarvae/day. The water in the tank is changed at a rate of 50%/day. After the change to the post larval form has occurred the water salinity is decreased to fresh water over a period of about 3 hours.

22.2 Nursery ponds

Nursery ponds are generally 0.1 ha in area with a depth of $\frac{1}{2}$ m. The postlarvae are stocked at a density of 125 000 per ha. The water flow is 600 litre/ha/day. The prawns stay in the nursery ponds for about 3 weeks, after which time they are transferred to the growout ponds.

22.3 Growout ponds

These vary in size between 0.2 and 2 ha, though sometimes even larger ponds are used. The ponds have a sloping bottom and an average depth of 1 m. A water flow of 200 to 600 litre/ha/day is maintained. The stocking rate varies depending on the intensity of cultivation required, the local conditions and availability of food; a common figure for stocking is 50 000 per ha.

The food used varies with availability and the intensity of farming. Animal and plant wastes and prepared feeds, including chicken feed, are used. The latter is the preferred food in many commercial farms and is fed at a rate of 6 g/ha/day initially, increasing to 37 kg/ha/day by the time the prawns reach market size. The amount of feed that can safely be given is related to the efficiency of the husbandry. Overfeeding leads to fouling of the water, which in turn leads to low oxygen levels, reduced growth rate of the prawn and high levels of mortality.

Fertilization is also used to stimulate growth of the algal mat on which the prawns feed; inorganic fertilizers up to levels of 25 kg/ha/month can be used; in some artisanal farms organic manure is used for the same purpose.

22.4   Harvesting

The prawns can be selectively harvested during the growth period by using a seine net with a stretched mesh size of 3–5 cm. Total harvest of the pond is completed by seine netting followed by draining the pond and hand gathering the crop.

22.5   Production

Production levels vary between 600 and 3000 kg/ha/year, depending on the type of farming practised. Market sized prawns of 30–40 g weight can be produced in 5–6 months.

Chapter 23
Oysters

The most important molluscs farmed, chiefly oysters and mussels, belong to the class Bivalvia. These species are extremely efficient converters of the primary production of the sea into animal protein. This is possible because they filter vast quantities of sea water through their feeding mechanisms and extract 90% of the plankton in the sea water which forms their food. These animals are capable of growing in high concentrations on small areas of substratum. In this way mussels can achieve the highest protein production per m² of farm area of any farm animal. True comparisons with terrestrial animals cannot, however, be made because of the enormous 'field' over which the bivalves feed. Experiments have shown, for instance, that a small oyster can pump 2–3 litres of water through its gills every hour, and as the tidal exchange of water over the farm areas is constant the animals filter off the productivity from vast areas of sea.

The oysters currently farmed belong to the genera *Ostrea* and *Crassostrea*:

Ostrea edulis	European oyster
O. lurida	Pacific oyster of North America
O. chilensis	South American oyster
Crassostrea angulata	Portuguese oyster
C. commercialis	Australian oyster
C. virginica	American oyster (USA)
C. gigas	Pacific oyster
C. rhizophorae	Mangrove oyster from South America

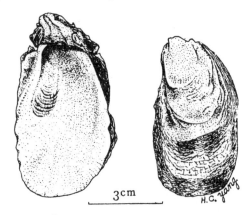

Fig. 23.1 Pacific oyster (*Crassostrea gigas*) (from Chen).

23.1 Culture

At present the culture of oysters covers a whole range of sophistication from the production of seed in hatcheries to the low-level management of wild stocks.

Hatcheries

O. edulis, *C. gigas* and *C. virginica* are all produced in hatcheries on a commercial scale. Ripe adult oysters are allowed to spawn under cover in tanks of running sea water. The density of the adults is four per litre. The larvae or eggs, depending on the species, are collected from the water in the tanks and placed in large plastic rearing tanks at a density of 1500 per litre. The tanks are aerated at a rate of 1 litre of air/hour for every 500 larvae. The major cause of death of the larvae is attacks from water-borne bacteria. To prevent these losses the water is treated with antibiotics as a standard procedure, a concentration of 50 IU of penicillin G and 0.05 mg of streptomycin sulphate per ml being used.

The larvae are fed with very small unicellular algae such as *Isochrysis galbana* and *Tetraselmis suecica*. These algae, which are about the size of a human red blood corpuscle, are produced in a separate culture system and added as required to the larval rearing tanks to maintain a concentration of 55 algal cells per microlitre of water.

When the larvae metamorphose and settle out as spat they are collected in the tanks on pieces of black PVC or on shells. The collectors, with the attached spat, are then placed in fibreglass tanks filled with sea water enriched with *Isochrysis* and *Tetraselmis*. The water is changed at weekly intervals and after about 6 weeks the collectors, with the attached spat, are placed outside in containers supplied with running seawater. When the spat reach a size of 1–2 cm they are detached from the collectors and placed on mesh trays at a density of 2 per cm² of surface area. Ten such trays of 45 × 23 cm in area require a flow of sea water of 1 litre/minute. The spat feed on the plankton which reaches them in the sea water. When they reach a size of 1 g, after about 6 months, depending on temperature and food abundance, they are ready for planting out on the growing beds.

Wild seed

Wild seed are collected as spat produced by oyster stock living in the collection region. The spat are collected on different types of material according to region. In Europe lime-covered roofing tiles are often used. The lime coating is to allow for easy removal of the developed spat. Various other materials are used, ranging from tar-covered sticks in Australia to oyster shells in the USA. The principal criterion for a successful collector is that the spat will settle on it and that they can be easily removed without damage.

23.2 Farming procedures

In Europe the spat settle in the summer and by the following spring they number about 1000 to the kg. They are then removed from the collectors and placed in trays which are stacked in threes in sheltered areas of the coastal water. The oysters are ready to be planted out the following summer.

23.3   Growout

Various growout systems are used. In many countries the young oysters are merely laid on selected areas of sea-bed in shallow coastal waters. In Australia the oysters are grown on lattices of tarred poles on which the spat were collected. These lattices are raised from the sea-bed on posts. The oysters are thinned out on the poles as they grow. In Japan all forms of culture are used: on the sea-bed, on racks raised from the sea-bed, on poles and on ropes stretched between poles.

23.4   Tropical oysters

In recent times there has been a growing interest in the potential for farming various species of oysters which grow in mangrove swamps. These are called mangrove oysters and include C. *tulipa*, C. *braziliana* and C. *belcherii*.

Chapter 24
Mussels

Mussels, which belong to the genus *Mytilus*, are distributed over the sea shores of the world. The genus contains many species, of which the following are farmed:

M. edulis	Europe
M. galloprovincialis	Europe
M. edulis chilensis	Chile and New Zealand
M. smaragdinus	Philippines
M. grayanus	India
M. viridis	India
M. edulis	Australia
Trichomya hirsuta	Australia

24.1   Seed production

Seed production depends entirely on the settlement of wild seed. The seed may be collected by hand from the rocks in which it settles naturally. Alternatively, poles are driven into the sea-bed, or sticks are placed near the mussel beds, or collectors are hung from floating systems to provide additional substratum on which the seed can settle.

24.2   Growout

The simplest form of growout system used is merely the distribution of the collected seed on suitable areas of sea-bed where the farmer can control the harvesting. A layer of mussels on the sea-bed is, however, subject to various disadvantages: the mussels are open to attacks by their natural predators which

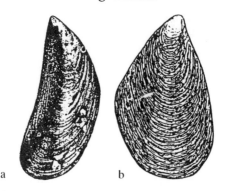

Fig. 24.1 (a) The blue mussel (*Mytilus edulis*); (b) The Mediterranean mussel (*Mytilus galloprovincialis*) (from Pillay: *Aquaculture: Principles and Practices*).

creep along the sea-bed; their access to pure sea water containing a high level of their natural food is limited by the reduced water flows along the sea-bed and by the higher turbidity of the water. These conditions can be improved by raising the mussels off the bed on ropes and poles. This type of improvement reaches its highest level in Spain where mussels are grown on long ropes suspended from rafts which are anchored in sheltered areas of sea. The Spanish mussels are thus separated from their predators and exposed to abundant clean plankton-rich sea water.

Mussel farming involves a minimum of husbandry work, which consists chiefly of thinning out the mussels as they grow to ensure that each one gets adequate access to the new sea water. The mussels which are removed in the thinning process are reattached to the ropes or poles by placing them in netting tubes or binding them on with netting bandages. As the young mussels grow they pass through the mesh and attach themselves on the outside. As they are being thinned any predators such as starfish are removed.

Very high levels of productivity are achieved by the raft system where ropes, perhaps 10 m long, are attached to the rafts. With these ropes productions of 600 000 kg/ha per year are possible, giving a yield of 300 000 kg of meat after shelling. Production on poles varies around 4500 kg/ha/year, yielding 2250 kg of meat.

Chapter 25
Scallops

Recent years have seen an increasing development in the farming of scallops. The following species are among a number now farmed, some more extensively than others:

Pactinopecten yessoensis	Japan
Argopecten irradians	China
Argopecten irradians	North America
Chlamys islandica	Iceland
Pecten maximus,	Europe
Chlamys opercularis	

The farming of scallops is similar to the farming of mussels. Natural spat are collected and grown on to market size, either suspended in cages or on ropes, or on the bed of the sea. The animals feed naturally from the water. Seed suppy is now augmented by hatchery techniques in a number of species.

Fig. 25.1 Rope culture of scallops (from Hardy: *Scallop Farming*).

Bibliography

Bardach, J.E., Ryther, J.H. & McLarney, W.O. 1972. *Aquaculture: the Farming and Husbandry of Freshwater and Marine Organisms*. Wiley-Interscience.

Beveridge, M. 1990. *Cage Aquaculture*. Fishing News Books.

Blakely, D.R. & Hrusa, C.T. 1990. *Inland Aquaculture Development Handbook*. Fishing News Books.

Bourne, N., Hodgson, C.A. & Whyte, J.N.C. 1989. A manual for scallop culture in British Columbia. *Can. Tech. Rep. Fish. Aquat. Sci.*, **1694**.

Bryant, P., Jauncey, K. & Atack, T. 1980. *Backyard Fish Farming*. Prism Press.

Chen, L. 1990. *Aquaculture in Taiwan*. Fishing News Books.

Edwards, D.J. 1978. *Salmon and Trout Farming in Norway*. Fishing News Books.

FAO. 1985. *Soil and Freshwater Fish Culture*. FAO Training Series 6, FAO, Rome.

FAO. 1981. *Water for Freshwater Fish Culture*. FAO Training Series 4, FAO, Rome.

Fallu, R. 1991. *Abalone Farming*. Fishing News Books.

Fishelson, L. & Yaron, Z. 1983. *Proceedings of an International Symposium on Tilapia in Aquaculture*. Tel Aviv University, Israel.

Hanson, J.A. & Goodwin, H.L. 1977. *Shrimp and Prawn Farming in the Western Hemisphere*. Dowden, Hutchinson and Ross Inc.

Hardy, D. 1991. *Scallop Farming*. Fishing News Books.

Harvey, B.J. & Hoar, W.S. 1979. *The Theory and Practice of Induced Breeding in Fish*. International Development Center, Ottawa, Ontario.

Horvath, L., Tamas, G. & Seagrave, C. 1992. *Carp and Pond Fish Culture*. Fishing News Books.

Imai, T. 1977. *Aquaculture in Shallow Seas*. A.A. Balkema.

Korringa, P. 1976. *Farming Marine Organisms Low in the Food Chain*. Elsevier Scientific Publishing Company.

Korringa, P. 1976. *Farming the Cupped Oysters of the Genus Crassostrea*. Elsevier Scientific Publishing Company.

Korringa, P. 1976. *Farming the Flat Oysters of the Genus Ostrea.* Elsevier Scientific Publishing Company.

Lee, D. & Wickins, D. 1992. *Crustacean Farming.* Blackwell Scientific Publications.

Leitritz, E. & Lewis, R.C. 1980. *Trout and Salmon Culture (Hatchery Methods)* California Fish Bulletin Number 164, Agricultural Sciences Publications, University of California.

Lutz, R. 1980. *Mussel Culture and Harvest, a North American Perspective. Developments in Aquaculture and Fisheries Sciences.* Elsevier Scientific Publishing Company.

Morton, J.E. 1967. *Molluscs.* Hutchinson.

Muir, J.F. & Roberts, R.J. (eds) 1982. *Recent Advances in Aquaculture.* Croom Helm.

Nash, C.E. & Shehadeh, Z.H. (eds) 1980. Review of Breeding and Propagation Techniques for Grey Mullet, *Mugil cephalus L. ICLARM Studies and Reviews* 3, International Center for Living Aquatic Resources Management, Manila.

New, M.B. & Sigholka, S. 1982. Freshwater Prawn Farming. A Manual for the Culture of *Macrobrachium rosenbergii. FAO Fisheries Technical Paper* 225.

Pillay, T.V.R. 1991. *Aquaculture: Principles and Practices.* Fishing News Books.

Pillay, T.V.R. 1992. *Aquaculture and the Environment.* Fishing News Books.

Pillay, T.V.R. & Dill, W.A. (eds) 1979. *Advances in Aquaculture.* Fishing News Books.

Shepherd, C.J. & Bromage, N. 1988. *Intensive Fish Farming.* Blackwell Scientific Publications.

Styczynska-Jurewicz, E., Backiel, T., Jaspers, E. & Persoone, G. 1979. *Cultivation of Fish Fry and its Live Food, Proceedings of a Conference September 23–8 1977 at Szymbark, Poland.* European Mariculture Society Special Publication No. 4.

Usui, A. 1991. *Eel Culture.* Second Edition. Fishing News Books.

Walne, P.R. 1979. *Culture of Bivalve Molluscs.* Second Edition. Fishing News Books.

Index

Books published by
Fishing News Books

Free catalogue available on request from Fishing News Books, Blackwell Scientific Publications Ltd, Osney Mead, Oxford OX2 0EL, England

Abalone farming
Abalone of the world
Advances in fish science and technology
Aquaculture and the environment
Aquaculture: principles and practices
Aquaculture in Taiwan
Aquaculture training manual
Aquatic weed control
Atlantic salmon: its future
Better angling with simple science
British freshwater fishes
Business management in fisheries and
aquaculture
Cage aquaculture
Calculations for fishing gear designs
Carp farming
Carp and pond fish
Catch effort sampling strategies
Commercial fishing methods
Control of fish quality
Crab and lobster fishing
The crayfish
Crustacean farming
Culture of bivalve molluscs
Design of small fishing vessels
Developments in electric fishing
Developments in fisheries research in Scotland
Echo sounding and sonar for fishing
The economics of salmon aquaculture
The edible crab and its fishery in British waters
Eel culture
Engineering, economics and fisheries
management
European inland water fish: a multilingual
catalogue
FAO catalogue of fishing gear designs
FAO catalogue of small scale fishing gear
Fibre ropes for fishing gear
Fish and shellfish farming in coastal waters
Fish catching methods of the world
Fisheries oceanography and ecology
Fisheries of Australia
Fisheries sonar
Fisherman's workbook
Fishermen's handbook
Fishery development experiences
Fishing and stock fluctuations
Fishing boats and their equipment
Fishing boats of the world 1
Fishing boats of the world 2
Fishing boats of the world 3
The fishing cadet's handbook
Fishing ports and markets
Fishing with electricity

Fishing with light
Freezing and irradiation of fish
Freshwater fisheries management
Glossary of UK fishing gear terms
Handbook of trout and salmon diseases
A history of marine fish culture in Europe and
North America
How to make and set nets
Inland aquaculture development handbook
Intensive fish farming
Introduction to fishery by-products
The law of aquaculture: the law relating to the
farming of fish and shellfish in Great Britain
A living from lobsters
The mackerel
Making and managing a trout lake
Managerial effectiveness in fisheries and
aquaculture
Marine fisheries ecosystem
Marine pollution and sea life
Marketing: a practical guide for fish farmers
Marketing in fisheries and aquaculture
Mending of fishing nets
Modern deep sea trawling gear
More Scottish fishing craft
Multilingual dictionary of fish and fish products
Navigation primer for fishermen
Netting materials for fishing gear
Net work exercises
Ocean forum
Pair trawling and pair seining
Pelagic and semi-pelagic trawling gear
Pelagic fish: the resource and its exploitation
Penaeid shrimps — their biology and
management
Planning of aquaculture development
Refrigeration of fishing vessels
Salmon and trout farming in Norway
Salmon farming handbook
Scallop and queen fisheries in the British Isles
Scallop farming
Seafood science and technology
Seine fishing
Squid jigging from small boats
Stability and trim of fishing vessels and other
small ships
The state of the marine environment
Study of the sea
Textbook of fish culture
Training fishermen at sea
Trends in fish utilization
Trout farming handbook
Trout farming manual
Tuna fishing with pole and line